Pocketbook for Technical and Professional Writers

Earl G. Bingham

STATE OF CALIFORNIA
DEPARTMENT OF WATER RESOURCES

and AMERICAN RIVER COLLEGE

WADSWORTH PUBLISHING COMPANY
Belmont, California
A Division of Wadsworth, Inc.

English Editor: Kevin Howat

Production Editor: Jeanne Heise

Designer: Patricia Dunbar

Copy Editor: Carolyn Davidson

Cover Designer: Patricia Dunbar

Printed in the United States of America

1 2 3 4 5 6 7 8 9 10—86 85 84 83 82

Library of Congress Cataloging in Publication Data

Bingham, Earl G.
Pocketbook for technical and professional writers.
Includes index.
1. English language—Rhetoric. 2. Technical writing. I. Title.
PE1408.B4959 808′.0666021 81–7628
ISBN 0–534–01004–0 AACR2

Contents

Preface

The communication problems analyzed in this book are particularly applicable to technical and business writing. The goal of technical and business writers should be to write so clearly that they cannot be misunderstood, and the *Pocketbook for Technical and Professional Writers* should help them reach that goal. Although the book is intended particularly for students of technical writing, professional writers and editors will also find it helpful.

The *Pocketbook* is divided into two parts, and a topical key, providing a table of contents for both Parts 1 and 2 appears on pages vii to xv. In Part 1, The Technical Writing Process, I have discussed clear writing from a reader's viewpoint. That is, I have emphasized the importance of *organizing* for readers and *writing* for readers. Part 1 stresses that writers must (1) use imagination to learn something about readers and what they will need to know; and (2) use accurate, specific words; construct clear, concise sentences; and create understandable paragraphs.

The articles in Part 2, A Technical Writer's Handbook, are in alphabetical order; topics range from the distinction between **a** and **an** to the difference between **your** and **you're.** Throughout the *Pocketbook,* I have used bold (**black**) type to indicate references to articles in the Handbook. Each article contains a specific example of the concept being defined or discussed; many of the examples are from the "real" world of science and government. For more than twenty years I have been a technical writer and editor in the aerospace industry and with an engineering organization of the State of California, and I have included many practical examples from both of those disciplines. During the past seventeen years I have been a writer/editor for the State of California; for the past nine years I have also taught at American River College, a combination enabling me to bring together the communication problems faced by professional writers and students of technical and business writing.

Using the Handbook is like using any alphabetical index. If your problem is capitalization, see **Capitalization** for examples of correct style. If you are uncertain about how to use abbreviations, see **Abbreviations** for examples of acceptable usage. If you are uncertain of the difference between **affect** and **effect, aggravate** and **annoy, discreet** and **discrete, principal** and **principle,** etc., see the Handbook, which clarifies those and some seventy other pairs of troublesome words. Are you concerned about the **metric system?** The Handbook contains a concise article and a handy metric conversion table. And if your sentences sometimes seem to wander off into outer darkness, **Sentence Types** and **Construction** will acquaint you with five important principles of sentence construction.

The Handbook also contains articles on **Application Letters and Résumés; Correspondence,** including several types of business letters; **Documentation,** with specific forms for footnotes, references, and bibliographies; **Oral Presentations; Proposals;** and **Reports. Graphic Aids** will acquaint you with the use of tables, charts, graphs, and photographs. You will also find discussions and examples of various "modes" of writing, including **Analysis, Cause and Effect, Classification, Comparison and Contrast, Definition, Description,** and **Process Analysis.**

As a self-help tool and reference source, this volume should be useful to all writers. On the other hand, it contains no magic formulas. Writers who would succeed must supply the imagination and industry required for clear technical prose. The *Pocketbook for Technical and Professional Writers* is intended to help them reach their goals.

I would like to thank: Richard J. Siciliano, Charles County Community College; R. S. Baker, Western Oregon State College; Thomas Munck, Blue Mountain Community College; Thomas Warren, Oklahoma State University; John Muller, Air Force Institute of Technology; Herman Estrin, New Jersey Institute of Technology; Nell Ann Pickett, Hinds Junior College; Wallace Coyle, Northeastern University; Joseph Saling, Ohio State University; Mitchell H. Jarosz, Delta College; James Drummond, C. S. Mott Community College; and C. Gilbert Storms, Miami University of Ohio, for their many constructive comments and helpful suggestions. Nor can I overlook the dedicated help of Kevin Howat and the entire staff at Wadsworth Publishing Company. Without their constant guidance, I could never have completed this volume.

Topical Key to the Text

Articles in the Handbook are in alphabetical order; topics range from the use of **a** and **an** to the distinction between **your** and **you're.** The articles, along with page numbers, are listed below in seven general categories: Planning and Research; Writing Style; Sentences; Grammar; Punctuation; Words; and Mechanics. Most topics are illustrated with examples of incorrect and correct usage.

Many of the Handbook articles are cross-referenced to related articles and to discussions in Part 1. For example, **Italic Type, Italicizing** is cross-referenced to **Underlining. Inflated Expressions** is cross-referenced to **Wise Words** in the Handbook and to *Watch for Redundancy and Circumlocution* in Part 1. An asterisk following a title or a page number indicates that the topic is discussed in Part 1.

Planning and Research

Writing Style

Sentences

Grammar

Punctuation

Words

General Articles

Words Often Confused

Mechanics

PART ONE

The Technical Writing Process

One of the most common barriers to effective communication is a writer's failure to consider the needs of readers. And a reader's needs can be considerable, particularly when writer and reader share no bond of knowledge regarding the subject. Writers too easily forget that they are at the end of the experience they are discussing but that many readers may be seeing it for the first time. Accordingly, if a writer's words are to be meaningful, the reader must have some knowledge of the subject matter. That is, the writer must use language the reader can understand.

Therefore, the first, and golden, rule of clear writing is always to consider the reader. Here is where a great many otherwise good writers go astray. They simply forget that their readers may be strangers to the subject matter.

In today's swiftly changing world, moreover, technical, business, and governmental writers must often discuss subjects of which their readers have only limited knowledge or perhaps none at all. An engineering writer, for example, may have to prepare a technical report to convince a board of directors that the corporation they oversee—whose business is manufacturing—should invest in costly new equipment. The writer is a technical expert; the board, however, comprises principally bankers, economists, and administrators, who are responsible for using the corporate funds wisely. What must the writer do to ensure that they understand the message? Or, a writer for a construction firm must prepare an important proposal to convince a governmental or private organization that her company can best build a dam or highway. The future of the company may depend on how well she writes the message.

Similarly, governmental writers must convince legislators that their states should invest in transportation, mental-health facilities, and environmental benefits, such as improved air and water quality. Legislators, who are responsible for administering the people's tax monies, are women and men with varying backgrounds. What will they need to know about the many subjects they must consider?

Still other writers must prepare instruction manuals to facilitate the maintenance and repair of costly, sophisticated machinery, such as computers, automobiles, aircraft, rockets, and televisions. The instructions must be understood by many different readers with varying backgrounds. Finally, letters and memorandums are vital to the success of almost every business and governmental organization.

For various reasons, then, hundreds of business and technical messages are written every day. Although their subjects and purposes vary, all written communications have this in common: they must be logically organized and clearly written for those who must understand them—the all-important and often-neglected readers.

Technical Writing

The technical communication problems discussed and analyzed in this volume apply to writing that presents facts, gives instructions, and interprets information—writing that is intended primarily to inform or instruct. This, as opposed to imaginative writing such as fiction and poetry, is known as expository writing, or exposition.

Technical writing is one form of expository writing, as are most business and governmental writing, all of the stories and informative articles in newspapers and magazines, and much other informative writing—history, biography, textbooks, instructional pamphlets, etc. With few exceptions, the papers and reports you write in the classroom will be expository. Thousands of expository papers and reports are written every day. The effectiveness of much of this writing is doubtful.

What is good expository writing? Actually, *good* writing is difficult to define, partly because the definition must necessarily be subjective and partly because the appropriateness of English usage is so highly dependent on the subject matter, the type of communication, and those who will read it. *Effective* exposition would be easier to define. Effective exposition is clear, concise, and considerate of those who must read it. It is writing that informs, primarily because a writer has organized it carefully and presented it clearly. Clarity is particularly essential in technical writing, for a very good reason. Readers who need technical information are busy readers, with no time to struggle through a poorly written report or memorandum trying to decipher a scrambled message.

Technical writers have a single purpose: to inform. To do so effectively, they must always consider their readers and how best to inform them.

An Analytical Task

One imaginative author of an instructive book on exposition calls this process *analytical writing;* that is, all writers should analyze their messages to be certain they are presenting well-organized, clear information. This author states that writers must make certain they are presenting well-organized paragraphs that unfold in logical sequence, rather than mere "catalogs of details" that readers must study and organize for themselves.[1]

What are some of the questions such an analysis might contain?

- Who am I writing for? How much do they know about the subject?
- Have I organized logically?
- Have I used words that most readers will recognize?
- Have I used concrete, rather than abstract, words?
- Have I used words correctly?
- Have I composed sentences that readers will understand the first time they read them?
- Have I divided the work into clear paragraphs?
- Have I used transitional words and phrases?

As you continue with this section and that on **Writing Style** (pp. 13–33), you will learn more about the points raised in the questions above.

Writer vs. Speaker

Take a moment to contrast a writer's task with that of a speaker. The latter has a number of advantages that writers do not enjoy. Speakers can use gestures and facial expressions and emphasize certain words to help put

[1]Thomas P. Johnson, *Analytical Writing* (New York: Harper & Row, Publishers, Inc., 1966), pp. 30–54.

across ideas. Writers, however, share none of those advantages. For them, words must work alone.

Above all, if listeners do not understand a speaker, they can ask for clarification. Writers, unfortunately, seldom have the advantage of answering follow-up questions: the readers may be in another part of the country; the writer may be unavailable to them for any of a dozen reasons. Even if a confused reader does get in touch, a great deal of time will have been wasted—and all because the writer failed to do the job correctly in the first place.

If you are one who must write—student, engineer, office supervisor, government employee—but you consider writing an impossible chore, your numbers are legion. But take heart! Just as natural laws govern the physical world, so do certain rules and principles exist as guides to clear technical and other professional writing. And you have already learned the first law of clear expression: always, but always, consider the reader.

Guidelines for Clear Technical Writing

There are, of course, many other guidelines to clear technical communication, and those are the subject of this book. Technical writers must organize facts, select words carefully, construct sentences that are both direct and grammatical, and weave their words and sentences into intelligible paragraphs. The first task in preparing any communication is its organization.

Organization

Effective organization of any written material usually results from careful planning. If your topic is short and uncomplicated, you may have only to jot down a few notes and select a logical order of presentation that readers will understand. A longer piece of writing, however—particularly when the subject is complex—will require more preliminary planning and probably an outline.

Any written presentation will consist of a number of individual facts and ideas. To get your ideas across, you must organize them to show their relative importance, how they relate to one another, and how they support your topic as a whole.

Assume that you must prepare a simple set of instructions on how to replace a defective wall switch, a household chore many people are generally familiar with. Suppose you wrote:

1. Remove the screws holding the defective switch in the wall box, after first removing the face plate.
2. Disconnect the wires from the defective switch and connect them to the same screws on the new switch.
3. Insert the new switch in the box and replace the face plate.
4. Be sure to turn the power off before you begin.

You could write instructions like those, but some of your readers might be in for a shock. In step 1, the facts are out of order; step 2 could confuse amateurs, particularly if they fail to note which wire they removed from each screw. Now, looking at the process from the reader's viewpoint, you could divide the instructions into four main steps with logical substeps beneath each main step:

A. Shut off the power at the circuit breaker or fuse box.
B. Remove the defective switch as follows:
    1. Remove the screws holding the face plate and remove the plate.
    2. Remove the screws holding the switch in the wall box and lift out the switch.
    3. Disconnect the black and white wires. Note that the black wire is connected to the dark (brass) screw, the white wire to the white screw.
C. Connect the new switch as follows:
    1. Attach the black wire to the brass screw and the white wire to the white screw.
    2. Before replacing the switch in the box, make certain the wires are attached snugly. Use a long-nosed pliers to shape each wire around the screw so that it cannot slip beneath the head.
D. Reinstall the switch, replace the face plate, and restore power at the breaker box.

Oversimplified as that example may be, it is now organized in a logical pattern that most readers could follow, even if this were their first attempt at electrical repair. And for those who have never looked inside a face plate, you could include a small sketch of the switch and attached wires:

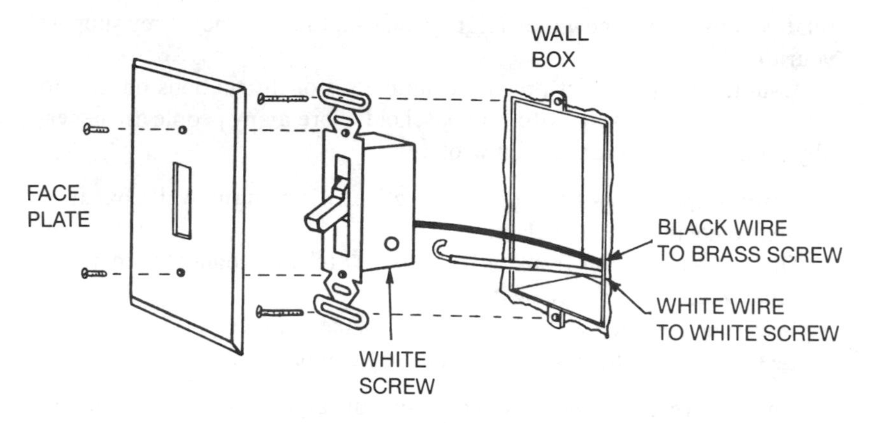

SINGLE-THROW SWITCH

Look again at the two sets of instructions. In the first, the writer apparently jotted down his or her thoughts without regard for the reader's skills, leaving a basic precaution as the final step and ignoring the importance of connecting the "hot" (black) wire to the dark screw. In the second set, however, the writer organized the instructions in correct sequence, that is, as the reader should carry them out.

The chances are that the second writer had thought about the audience carefully, realizing that some may never have done more than change a light bulb. Therefore, he or she covered every base and listed the facts in exact order. Have you ever listened to an explanation in which a speaker was forced to backtrack with a " . . . just a moment, I forgot to mention the . . ."? Nothing is more confusing than receiving jumbled facts in random order.

Methods of Organization

There are several methods of organizing information, not all of them equally effective. One way would be to arrange information randomly, as the cards are dealt from a shuffled deck. A random arrangement would be suitable for, say, a collection of jokes or a series of brief anecdotes, such as "Life in These United States," a regular feature of *Reader's Digest*. None of the items bears a direct relationship to the others, except that each presents similar, although unrelated, information. Each is complete in itself and independent of the others; therefore, a random arrangement is appropriate.

You could also arrange facts numerically or alphabetically. However, except for a parts list, a set of instructions, a glossary, a directory, or a bibliography, a numerical or alphabetical list could result in confusion. For instance, the parts of an air conditioner could be listed in alphabetical order or by the manufacturer's part numbers. A set of instructions on how to repair an air conditioner would undoubtedly be arranged as a series of numbered steps. An explanation of how an air conditioner functions, however, would not lend itself to an alphabetical or numerical arrangement. The "parts" of that discussion could best be presented in order of importance, which is a third method of arranging information.

In the third system, facts of first-order importance are ranked equally, and each main fact is divided into a number of subordinate facts. I call this system the *whole-to-the-parts* method of organization—dividing the overall subject into parts, or subtopics, in order of importance.

Dividing a topic into parts is a most logical system for organizing a discussion of any length, because it best shows the relationships among facts. When these relationships are clear, readers will be able to comprehend the information most readily and retain it in their long-term memories.

For instance, the number 431 would be easy to recognize and remember. On the other hand, the number 9614875685 would be difficult to recognize and almost impossible to remember. However, if the overall number were separated into logical groups, you would quickly recognize it as a telephone number: area code 961; exchange 487; number 5685. Now the number 961-487-5685 makes sense and would be relatively simple to recognize and recall.

The same principle can be applied to technical writing. By logically dividing and organizing a subject, you make it easier for readers to both grasp the essential facts and determine how those facts are related. To organize from the whole to the parts, then, you would divide the subject into several main parts, with subordinate parts grouped beneath each main part.

Use the Whole-to-the-Parts Method

The best way to explain the method is with an example. Assume that your specialty is sanitary engineering and that you must prepare a brief report on some aspect of waste water. You have researched the subject thoroughly and are prepared to begin. You could go about it in either of two ways: (1) you could simply begin writing without a formal plan and hope that what you wrote would make sense to readers, or (2) you could begin by asking yourself three important questions. The first one concerns the audience; the last two concern the topic.

1. *Who will read this report?* Assume that your readers will be beginning engineering students with little knowledge of the subject. This will help you decide how much you will have to tell them if they are to understand the report.
2. *What specific aspect of waste water am I writing about?* Obviously, the general topic *waste water* is too broad a subject for a brief report, and you would have to narrow it to manageable size. Let's assume that you decide to discuss the *treatment* of waste water.
3. *On what basis shall I divide my topic into subtopics?* Since there are three general treatment processes—primary, secondary, and tertiary (also called advanced)—these will become the three main divisions for your report.

State a Thesis (See **Thesis** in the Handbook)

You might try a tentative thesis statement covering the entire subject. For instance:

> The basic objective of waste water treatment is to remove or neutralize pollutants added to water during municipal, agricultural, and other uses, so that the water can be safely discharged or reused. There are three general levels of treatment: primary, secondary, and tertiary (often called advanced).

Don't worry about perfecting your opening statement at this point; you can revise it later. In fact, you may find it simpler to defer a general introduction until after you have written the body of the report. By then, you will know every part of the report thoroughly, and you will be able to frame a clear introduction.

On the other hand, a general thesis statement will help you decide how to organize your material. Thus, you will find it helpful to decide on a thesis before you begin organizing. The thesis answers the "what about" question and will suggest how you should develop the report. It will also remind you of the limits of your subject. As you continue to plan, keep your thesis at hand to remind you of the route you should be taking.

Brainstorm

To decide on the various parts, or subtopics, that you will need, it will be helpful to jot down all the relevant ideas you can think of. By jotting down ideas as they come to mind, you will soon have enough for an outline. It won't matter if you finally decide that you have too many or that some are irrelevant. You can easily discard the surplus and then begin to organize those you decide to retain. (See also **Brainstorming** in the Handbook, p. 64.)

Construct the Framework

A brief skeleton outline on waste water treatment might look like this:

The Treatment of Waste Water

General introduction
I. Primary treatment
    A. Specific objectives
    B. Treatment processes
        1. Conventional primary treatment
        2. Imhoff tank plant
        3. Stabilization ponds
    C. Disinfection following primary treatment
II. Secondary treatment
    A. Specific objectives
    B. Treatment processes
        1. Trickling filter plant
        2. Activated sludge plant
        3. Oxidation ponds
    C. Disinfection following secondary treatment
III. Tertiary (advanced) treatment
    A. Specific objectives
    B. Treatment processes
        1. Removal of nutrients
            a. Phosphorous removal by chemical precipitation
            b. Nitrogen removal by air stripping
            c. Settling ponds
            d. Activated carbon
        2. Removal of minerals (demineralization)

a. Electrodialysis
b. Reverse osmosis
c. Ion exchange
d. Distillation

Complete the Outline

After sketching a skeleton outline, you must decide on the exact details, and perhaps additional subpoints, for each of your headings. You will find it best to expand the outline in full detail, so that you don't overlook anything as you are writing. It is far better to spend a few hours on the outline than to discover, once you have completed the writing, that you have omitted something important. Preparing a detailed outline will also enable you to decide exactly what you will have to include if you are to inform your readers fully.

On the other hand, an outline should be flexible. That is, even after you have begun to write a draft, should you change your mind about certain facts or decide that some facts are out of order, you can go back to your outline to determine where those facts belong. You will usually find it easier to change an outline than to "undo" a completed text. Then, after reworking the outline, you can revise the text as needed.

Summary

The value of the whole-to-the-parts method and of an outline can be summarized as follows:

1. Dividing your subject will help you organize your thoughts and arrange your material in logical order. To organize the report on waste water treatment, you would divide the overall subject (the whole) into logical subdivisions (the parts).
2. The outline will help keep you on track as you write. As you discuss primary treatment, for example, the outline will remind you that you are discussing the first stage of waste water treatment, not the removal of nutrients or minerals, which is carried out in advanced, or tertiary, treatment.
3. The outline will remind you of all the important parts in your report.
4. The outline is a flexible blueprint to your report. You can change it easily as you see fit.
5. The outline will give you a place to start and a place to stop.

In a word, to organize efficiently, divide your subject into parts and use an outline as a builder uses a blueprint.

Writing Style

If you have carefully organized your message, its effectiveness will be determined by how you put your words on paper—in other words, by your writing style. Writing style cannot be learned, but it can be improved. That is, writing style is similar to personality—it is unique to every person. And just as you can improve your personality, you can enhance your writing style by conscientiously striving to

1. Say what you mean—clearly.
2. Choose words carefully.
3. Write concisely (but not too concisely).
4. Place modifiers where they belong.
5. Create effective sentences.
6. Create meaningful paragraphs.
7. Use transitional words and phrases.
8. Revise, at least once, everything you write.

Let's examine those points in some detail.

Say What You Mean—Clearly

Many sound ideas get lost in technical writing, or in any other exposition, because a writer has expressed them in dull, wordy, or involved style. Readers will have trouble following—and may even ignore—writing that is difficult to understand. Even a simple concept may sometimes be incomprehensible: *It is mandatory that the necessary action be implemented to facilitate prompt conclusion of this project* will probably require a second reading. *This project must be completed at once* says the same thing, and the meaning can be determined at a glance. (See also **Gobbledygook** in the Handbook.)

Here are a few other examples of verbosity and confusion that have crossed an editor's desk:

1. Never have I read such an esoteric discussion so clearly written.
2. The initial action is that of the actuators, which raise the missile into launching position.
3. The accumulated exposure, if held within these limits, will ensure that the radiation worker is not subjected to occupational hazards greater than those experienced by his industrial counterpart in any other industry which does not employ radioactive materials as an instrument with which to accomplish a productive task.
4. This report presents four alternative courses of action or inaction which can be taken to improve or possibly lose the many resources of the Sacramento–San Joaquin Delta.

The writers of each of those passages had something definite in mind but let their readers down. Without even trying, the writer of example 1 created the perfect paradox, or contradiction in terms. He meant, of course, *such a clear discussion of an esoteric subject.* The extra words in example 2 do nothing to clarify the meaning, which is simply, *First the actuators raise the missile into launching position.*

The author of example 3 had worked overtime. After a brief debate, he agreed to delete some three dozen words from the original statement. This brought out the meaning in far fewer words: *Such exposure limits materially reduce the hazard to radiation workers.* The writer of example 4 took just the opposite course and tried to skim by on too few words. He meant: *This report presents four alternative plans that will affect the future of the Sacramento–San Joaquin Delta. Three of the four would result in varying degrees of improvement. However, if present maintenance practices are continued, as under one alternative, we could actually lose many of the Delta's valuable resources.*

Many other amusing examples of confusion are available, but such writing is seldom entertaining to readers who must read important reports and memorandums every day. When the preceding examples were revised *with readers in mind,* the meaning of each became clear.

Choose Words Carefully

Clear writing often depends on careful selection of words. Technical writers must (1) avoid unusual or highly technical words, which may baffle readers; (2) avoid abstract words, which create vague ideas; and (3) use the correct word in a given situation.

Avoid Highly Technical Words

It is only natural that experts may want to write with dignity and demonstrate their knowledge of the subject. Unfortunately, when they use technical words, which are often meaningless to readers who lack their knowledge, they may only create an impression of pomposity.

For instance, if a geologist described a particular rock formation as *friable, argillaceous,* and *cemented by interstitial silt,* his or her colleagues might be impressed. Most nongeologists, however, would probably reach for lighter reading. What's more, all the unusual words in that passage could have been replaced, or at least explained, with more common words: *friable* means *brittle* or *readily crumbled; argillaceous* means *resembling or containing clay; interstitial* means *occurring in minute cracks, or interstices.* For that matter, *silt* could have been further defined as *fine mineral particles varying in size from sand to clay.*

A word to the wise: why use an unusual word if a simpler one will express your meaning? And if you must use technical words, explain them somehow. If your readers are acquainted with the subject, they won't mind a brief explanation. If the subject is new to them, however, they may be lost without it. Readers who enjoy being puzzled read mystery stories. Technical and business readers are busy readers who want a clear message the first time. (See also **Pompous Language** in the Handbook.)

Abstract and Concrete Words

Abstract words present ideas so far removed from direct experience that readers cannot determine what the words stand for. Concrete words refer to objects or ideas that readers can readily visualize. Such vague summarizing expressions as *good condition, corrective action,* and *extended period of time* will only puzzle a reader who needs specific information. For example, if you wrote, "Construction was stopped for several days because of unfavorable weather," readers would learn neither how the weather was unfavorable nor for how long. By changing your words to, "Construction was stopped for three days because of heavy rain," you would answer both questions.

Writers must be equally careful with inexact expressions such as *a few, several, reasonably, substantial,* and with other words that can have different meanings for different readers. If you were to write, "In recent years the population increased substantially," your audience would remain uninformed. Accordingly, if you were discussing a certain five-year period—say 1976 through 1980—and a population growth of 40 percent—from 200,000 to 280,000—you should say so instead of leaving your readers in a fog.

On the other hand, there is no law against introducing a discussion with a summary statement, providing you also include the details. You will often find it best to introduce a subject with a broad generalization, followed by specific examples to illustrate your point. For instance:

General: We plan to take several new fire precautions. *Now the details:* We will install fire doors, an automatic alarm system, and an automatic sprinkling system.

The obvious difference between the abstract word *precautions* and the three specific examples is that the former leaves only a fuzzy impression in a reader's mind, whereas the three examples paint specific pictures that readers can readily visualize.

Use the Correct Word

Writers must be especially careful with words that sound alike but are spelled differently (see **Homonyms** in the Handbook). They must also watch out for words that appear to be synonymous but actually are not.

In the first category are such homonyms as *affect* and *effect, compliment* and *complement, discrete* and *discreet.* Such words are often misused, sometimes because writers don't know the difference and sometimes because they fail to check their work. The misuse of words may not destroy a writer's meaning, but it can damage his or her reputation when readers discover the error, as many readers inevitably will.

In the second category are words that closely resemble other words with different meanings. Take, for example, the words *different* and *indifferent.* They may appear related, but their meanings are quite unrelated. This statement appeared in a recent report:

Because of indifferences between the contractor and the state [of California], the work was delayed.

The writer meant, of course, difference of opinion, not lack of concern. The word he really wanted was *disagreement* or *dispute.*

Here is a student's introduction to a theme on communication:

In today's world, communication is omnipotent.

The student later revealed that he meant all-important. Omnipotent means all-powerful, a trait usually ascribed only to God.

This sentence appeared in the first draft of a brochure describing a popular recreation area:

Campers arrive in droves because the climate is most salubrious.

Challenged by an editor, the writer insisted that *salubrious* truly described the area and is "a marvelous word."

"It means *conducive to health*," the editor countered. "Is that what you mean?"

"No," the writer confessed. "I should have said *salutary*."

"That also means conducive to health. What did you mean?"

The writer now surrendered. "Make it," he said, " 'the climate is pleasant.' "

Errors like those may cause considerable confusion and delay while readers try to determine a writer's meaning. At best, **malapropisms** will impair a writer's image, particularly if the reader enjoys a good laugh at the writer's expense. The best cure for this problem is the dictionary. (See also **Malapropisms** and **Spelling** in the Handbook.)

Write Concisely (But Not Too Concisely)

You are writing concisely when you use a minimum number of words to make a point. Conciseness, however, is a two-way street. Look first at the problem caused by **inflated expressions.**

Watch for Redundancy and Circumlocution

Writers often clutter their work by using more words than necessary or by inadvertently repeating an idea. Technically, the use of two or more words for one is called redundancy; writing that goes around in circles with excessive words and phrases is called circumlocution. (For an analysis of various types of circumlocution, see **Inflated Expressions** in the Handbook.)

After the 1979 accident in the Three Mile Island nuclear power plant, the United States Nuclear Regulatory Commission reported:

It would be prudent to consider expeditiously the provision of instrumentation that would provide an unambiguous indication of the level of fluid in the reactor vessel.

Translated: We need more accurate measuring instruments. (See also **Gobbledygook** in the Handbook.)

Not long ago, I read this bit of gardening advice:

During early spring, *following winter,* perennials should be *pruned back so as to* encourage the *putting out of new growth.* (Italics mine)

"When else does spring occur?" I muttered. "I could hardly prune them forward, and where else could the new growth go but out?" Here is a simpler version: Perennials should be pruned every spring to encourage new growth.

Writers who use two or more words for one or who create inflated expressions serve no purpose but to increase a reader's work. For instance, an item that is red *in color* is only red; a situation that is controversial *in nature* is only controversial. When you *put on a demonstration,* you merely demonstrate, and to *provide a reduction* is only to reduce. What's more, a charge *in the amount of* $3.50 remains $3.50.

Most of us are guilty of **circumlocution** in conversation, and the extra words we use often creep into our writing. Remember, inflated writing may obscure your meaning or at least cause readers to search for it. Even if they do understand your message, you will surely increase their reading time.

Take this example (from a student paper): *In the area of advantages as to the use of oil there are three.* It may take you a few minutes to determine that the writer meant, simply, *the use of oil has three advantages.* Here is another example of pure inflation: *Learning to land by means of instruments is a task that must be mastered by all student pilots.* This simply means: *All student pilots must practice instrument landings.* In each of those inflated sentences, fourteen words were reduced to seven, a 100 percent reduction in word count and an equivalent increase in readability. Moreover, such "factoring" may reduce a reader's time some 200 percent, especially when the subject matter is complex.

The real problem with inflated expressions is the number of them that writers manage to accumulate in adjacent phrases and sentences. By itself, a superfluous word or two might never be noticed. But multiply the extra words in one sentence by the number of sentences in a ten-page memorandum or a fifty-page report, and you will see that the possibilities for overwhelming readers become staggering. (See also **Inflated Expressions** and **Prepositional Phrases** in the Handbook.)

Avoid Superconcise Writing

Writers sometimes create a different type of confusion by using too few words. This problem arises from the use of noun **modifiers** to the exclusion of prepositions, resulting in what one author calls "superconcentrated style."[2] Faced with prose that is too concise, readers are often forced to separate the modifiers to determine the relationship between the nouns. I remember spending about ten minutes on the following sentence (written by a metallurgist who was discussing a test of steel plate):

The data will permit selection of the required plate material strength level and chemical composition.

At first I thought I was reading about selection of three things (plate material, strength level, and chemical composition) and that a pair of commas was missing. Finally, I learned what this economy-minded writer had meant:

The data will permit selection of the required strength level and chemical composition *for the plate material.*

Now it seems rather elementary; the use of one **prepositional phrase** (*for the plate material*) brought out the writer's meaning. But the writer should have clarified the sentence for me: writers should never deceive readers and compel them to dissect sentences.

Here is another superconcise sentence:

They developed a chemical precipitation nitrogen and phosphorous removal process.

Once again, the reader runs into five nouns too close together for comprehension. What did "they" develop?

a chemical precipitation process *for removal of nitrogen and phosphorous*

The moral of the story: don't be overly loquacious, but don't try to save too many words.

[2]Theodore A. Sherman and Simon S. Johnson, *Modern Technical Writing,* 3rd ed. (Englewood Cliffs, N.J.: Prentice-Hall, Inc., 1975), p. 24.

Be Careful with Modifiers

The usual advice about sentences is to keep them short to avoid confusion. Although long sentences are sometimes difficult to follow, the real trick is to shorten the distance between closely related sentence elements. Skillful writers do this by placing **modifiers** close to the modified words and phrases. Moreover, writers who use modifiers carefully can use **compound** and **complex sentences** at will. On the other hand, even a short sentence carelessly thrown together can cause confusion. For example:

1. The rocket propellant is contained in a cylindrical thrust chamber, which burns at approximately 5,000° F. (The chamber burns?)
2. The records now include all test reports for engines received from the new test facility. (Engines received from the test facility?)

Each of those brief statements is confusing because the writer failed to consider word order. And in each case, the remedy is the same: shorten the distance between the modifier and the modified element. In other words, keep modifiers close to the words they modify.

The writer of example 1 (above) burned up the rocket instead of the propellant. What did he mean?

The propellant, which burns at 5,000° F, is contained in a cylindrical thrust chamber.

Readers of example 2 might have wondered what exactly did arrive from the test facility. The actual meaning:

The records now include all engine-test reports from the new test facility.

Those two examples should illustrate how easily writers can mislead readers without even trying. You will never be guilty of misleading if you remember that modifiers should be as close as possible to the modified word or phrase. Otherwise, readers will be compelled to stop and move sentence elements around until they have determined what words go with what. And that is the writer's job.

Sometimes, for the sake of *style,* a writer will deliberately divide the **subject and predicate** by inserting a modifier between them. Often this can be just enough to throw readers off the track. For example:

No single number, in all of its compounds, can be used to represent the radius of chlorine.

In that statement, the modifier *in all of its compounds* appears to modify *number,* and grammatically it does so. Actually, the writer meant it to modify *chlorine,* and the modifying phrase is definitely out of place. As revised, the statement is easily understood:

No single number can be used to represent the radius of chlorine in all of its compounds.

Still another type of misplaced modifier is the squinting modifier, so called because it "looks both ways" in a sentence. The squinting modifier shows up between two clauses, forcing readers to decide which clause it refers to:

Although the Oroville area has long been considered a region of low seismic activity, prior to 1975, 58 earthquakes were recorded within a 37-mile radius of the dam.

To make certain no readers were deceived, the writer changed the construction to two sentences:

The Oroville area has long been considered a region of low seismic activity. Before 1975, however, 58 earthquakes were recorded within a 37-mile radius of the dam.

Misplaced modifiers are probably the all-time champions of confusion and ambiguity. Watch your modifiers closely, and don't make readers work overtime. (See also **Dangling Constructions** in the Handbook.)

Control Your Sentences

How long should a sentence be? Some writers reduce this concept to a formula, but I would be among the first to demur. As you will soon see, short sentences do not necessarily make writing crystal clear. Actually, there is no magic formula for determining sentence length. A general rule might be that highly complex subjects are more easily understood in short doses. For instance, most of us would probably have to read the following sentence at least twice:

The satisfaction, or utility, that a person derives from spending income must be measured against the disutility of earning it, and, therefore, in a free society every person will choose the type and amount of work that returns the greatest satisfaction, and except to supply material wants, the average individual will not remain long at frustrating and unsatisfying employment.

That sixty-four-word sentence, a discussion of abstract economic theory, simply presents too many ideas too quickly, and its stringy construction hides the relationships between the ideas. It really should be three sentences:

The satisfaction, or utility, that a person derives from spending income must be measured against the disutility of earning it. Therefore, in a free society, every person will choose the type and amount of work that returns the greatest satisfaction. And, except to supply material wants, the average individual will not remain long at frustrating and unsatisfying employment.

Now the **paragraph** is easier to follow. It opens with a **topic sentence** followed by two statements in support of the topic.

Now read this sentence about the weather:

Violent changes in the weather frequently lead to huge losses of property and losses of life as well, and in the United States alone, annual losses from hurricanes average 250 million dollars, in addition to which hurricanes have killed more than 15,000 Americans since 1900.

Again, the sentence should have been three sentences:

Violent changes in the weather frequently lead to losses of property and income, and sometimes to losses of life. In the United States alone, annual losses from hurricanes, for example, average 250 million dollars. Moreover, hurricanes have killed some 15,000 Americans since 1900.

On the other hand, with less complex material, sentences may contain thirty, forty, and even more words and still be perfectly understandable, particularly when the writer expresses parallel thoughts in parallel form and when the words contribute to a single idea. (See **Sentence Types and Construction** in the Handbook.) Consider this descriptive sentence:

The South Lahontan Basin, a part of the Great Basin, is characterized by numerous enclosed sinks and dry lakes, and by the greatest extremes in elevation in the coterminous United States, ranging from 282 feet below sea level in Death Valley to 14,495 feet at the peak of Mt. Whitney.

Although that sentence contains fifty words, the words add up to a single idea—a simple physical description—and the descriptive phrases are in parallel form.

The question of sentence length demands careful attention. Earlier I mentioned the author who suggests that expository writing might well be called analytical writing: that is, writers must analyze their work for clarity and readability. An important part of such analysis will concern the length of sentences.

To repeat: whenever the topic is complex, the sentence structure should probably be simple. With complex material, several short sentences will usually be clearer than one long sentence, because readers need absorb only one idea at a time. The period following each short sentence gives the reader a breathing spell.

Avoid Short, Choppy Sentences (Primer Style)

A writer can also confuse readers with too many short, choppy sentences, sometimes called primer style. (Remember "Dick and Jane"?) Faced with a string of short sentences, a reader runs into another problem: the brain literally gets ahead of the eyes. The eyes must stop at each period, but the brain wants to keep going. What's more, such constructions are often repetitious; a writer often follows a short sentence with a second sentence restating part of the first.

In fact, had the writer of the preceding descriptive sentence divided it into several short sentences, he would have impaired the flow of the descriptive effect and ended up with unneeded repetition. For instance:

> The South Lahontan region is part of the Great Basin. *The region* is characterized by numerous enclosed sinks and dry lakes. *It also has* the greatest extremes in elevation in the coterminous United States. *These extremes* range from 282 feet below sea level in Death Valley to 14,495 feet at the peak of Mt. Whitney.

Note the choppiness and repetitious rhythm in that version. Such writing is not only uninteresting, it is also difficult to absorb. In the original version, the writer used **apposition** ("a part of the Great Basin"), two parallel **prepositional phrases** ("by numerous . . . and by the greatest extremes . . ."), and a **subordinate clause** that lets the reader see the extremes in elevation. Despite its length, the statement is easy to read because its parts are parallel.

Here are a few more examples of choppy writing:

> Medical studies have shown that certain diseases are affected by changes in the weather. Some of these diseases are arthritis, rheumatism, asthma, and certain heart ailments.

The repetition can be removed easily by using the active voice and combining the two simple sentences into one:

Medical studies have shown that changes in the weather affect certain diseases, such as arthritis, rheumatism, asthma, and certain heart ailments.

Combine Short Sentences

Consider these four simple sentences, which are replete with repetition:

Salt with finely ground crystals can be processed. The crystals are only 0.0004 inches in diameter. However, this salt is expensive to produce. The cost of this salt is about $3.00 per pound.

Once again, the repetition can be eliminated easily:

Salt with finely ground crystals only 0.0004 inches in diameter can be processed, but the cost is about $3.00 per pound.

You can also combine short sentences by using one or more of the constructions discussed in these Handbook articles: **Absolute Phrase; Appositive; Compound Predicate; Participle; Restrictive and Nonrestrictive Modifiers;** and **Sentence Types and Construction.** To illustrate briefly, here are several examples of how you could use those constructions to combine two ideas. First, consider these two short sentences:

I finished my chemistry report. Then I went to the park.

The two ideas appear in two simple sentences, each with equal weight—two independent clauses (see **Clauses and Phrases** in the Handbook). If you wanted to emphasize each idea equally, two independent clauses might be appropriate. For instance, you might be answering the question, "What did you do yesterday afternoon?" You could also combine the ideas and present them in a variety of alternative constructions:

1. **Absolute Phrase** (Handbook, p. 45): My chemistry report finished, I went to the park.
2. **Appositive** (Handbook, p. 59): I finished my homework, a chemistry report, and went to the park.
3. **Compound Predicate** (Handbook, p. 80): I finished my chemistry report and went to the park.
4. **Complex Sentence** (Handbook, p. 243): After I finished my chemistry report, I went to the park.
5. **Compound Sentence** (Handbook, p. 243): First I finished my chemistry report, and then I went to the park.

6. **Participial Phrase** (Handbook, p. 187): After finishing my chemistry report, I went to the park.
7. **Nonrestrictive Modifier** (Handbook, p. 240): I finished my homework, which was a chemistry report, and went to the park.

How you combine ideas will depend on what you consider most important—what you want to stress. Normally, the independent clause(s) will stress the most important idea(s). In the preceding list, sentences, 1, 4, and 6 emphasize going to the park; sentences 2 and 7 tend to subordinate the park and stress the report; sentences 3 and 5 present both ideas with equal emphasis.

Vary the Rhythm

By now you may have decided that sentence length is indeed a matter of a writer's judgment and analysis. You might also consider the old adage about variety and the spice of life. Just as with music, constant repetition of the same sentence rhythm will soon turn readers off. The best solution to that problem is to make some sentences short, some medium length, and others somewhat longer.

Thus far, we have looked at modifiers, sentence length, and sentence variety; there is much more. For five other important principles of sentence construction, see **Sentence Types and Construction** in the Handbook.

Create Meaningful Paragraphs

A paragraph may be defined as a topic unit, that is, a group of related sentences unified by a topic sentence or controlling idea. Most paragraphs should open with a controlling idea, or central thought, expressed in a topic sentence. Stated another way, the topic sentence introduces the topic and ties together the supporting ideas.

Think of the topic sentence as a miniature thesis—that is, a general statement of what the paragraph is all about, what the reader can expect to find in the paragraph (see **Thesis** in the Handbook). For example, in the following paragraph the topic sentence is italicized. The remaining sentences directly support the topic statement:

Employment in California is expected to grow steadily over the next twenty years, although job categories will change somewhat. New, sophisticated farm machinery will continue to reduce on-farm employment, while jobs in other resource-based industries—mining, forestry, and fisheries—will increase slightly. Manufacturing jobs will increase by some 700,000, but, in response to the demands of an expanding and affluent society, service and government jobs will show even larger gains.

Note how the three sentences following the topic sentence support the controlling idea:

Topic sentence (controlling idea):

Over the next twenty years, employment will increase; job categories will change.

Supporting ideas

1. Farm employment will decrease.
2. Jobs in mining, forestry, and fisheries will increase slightly.
3. Manufacturing jobs will increase by 700,000.
4. Service and government jobs will show the greatest gains.

To change the topic, or to introduce a new idea about the topic, a writer should begin a new paragraph. Otherwise, readers will be into the new topic or idea before they realize the subject has shifted. At best, some will be able to decipher the change for themselves; at worst, many will be hopelessly lost. Note how the subject shifts in the following two paragraphs. In both, the topic sentence is italicized:

Opening paragraph

California's natural water supplies are derived from an average annual precipitation of 200 million acre-feet—the equivalent of 65 trillion gallons. About 65 percent of that precipitation is consumed through evaporation and transpiration by trees, plants, and other vegetation. The remaining 35 percent comprises the state's annual runoff of 71 million acre-feet.

New idea

The wide disparity in available runoff, both from year to year and between different parts of the state, has created the need for storage and conveyance of surface water. Ironically, the largest amounts of runoff are available in areas with the fewest people, specifically the north coast and the Sacramento Basin. Accordingly, as California has grown, its surface-water systems have been expanded to large-scale transfer systems that transport water through much of the state.

Because the writer recognized the separate ideas, he divided them into separate paragraphs, the first explaining the source of water (idea 1) and the second introducing a natural problem (idea 2). Thus, a change in paragraphs helps readers recognize a shift in ideas.

At the other extreme, to avoid the risk of creating overly long paragraphs, a writer might be tempted to make every sentence a separate paragraph. This would be equally unwise, because readers would then have to struggle to connect the ideas that support each topic sentence.

The principal paragraphing problems most writers face are (1) excessive paragraph length, (2) the need for the transition within paragraphs, and (3) the need for transition between paragraphs.

Consider the Length of Paragraphs

Frequently, the topic sentence—the controlling idea—will be a "compound"; that is, it may consist of several parts. In such instances, each part may require a separate paragraph. By way of illustration, consider the following paragraph, part of an explanation of how a city with hard water supplies might benefit from a central water-softening plant. This particular paragraph was intended as a general explanation of economic benefits. (The topic sentence is italicized.)

> *Benefits may be classified as primary or secondary, and tangible or intangible.* Primary benefits result from an increase in the value of products or services, a reduction in costs, or a reduction in damage or losses. Examples of primary benefits that would result from installation of a water-softening plant are the increased life of water-using appliances and plumbing fixtures, reduced requirements for soap and washing powders, and reduced need for bottled water and home water softeners. Secondary benefits are indirect benefits, such as improved growing conditions for lawns and plants, which might improve the appearance of the community, attract new residents, and thus increase economic activity. Tangible benefits are those that can be calculated in dollars, such as the cost of bottled water. Intangible benefits, such as an improved community appearance, cannot be readily expressed in monetary terms.

The difficulty for most readers of that dense paragraph is caused by the four different parts of the controlling idea: primary, secondary, tangible, and intangible. Unless you happen to be an economist, you would probably have to pause at each part and separate the shift in thought. For those unfamiliar with the language of economics, the four different, although closely related, parts of the controlling idea are simply too close together for easy comprehension.

Now, let's divide that fact-packed paragraph into three paragraphs and see what happens. We can open with the same controlling idea; then, if we separate its four parts into smaller, less-dense topic units, readers should be able to absorb the information more quickly. Note that the second and third paragraphs now have their own topic statements, further explaining the original controlling idea.

Benefits may be classified as primary or secondary, and tangible or intangible. Primary benefits result from increased values of products or services, reduced damage or losses, and reduced costs. Examples of primary benefits that would result from installation of a water-softening plant are the increased life of water-softening appliances and plumbing fixtures, reduced requirements for soap and washing powders, and reduced need for bottled water and home water softeners.

Secondary benefits are the indirect benefits that can be attributed to the proposed new softening plant. They include such things as improved growing conditions for lawns and plants, which might improve the appearance of the community, attract new residents, and thus increase economic activity.

Both primary benefits and secondary benefits may be tangible or intangible. Tangible benefits are those that can be calculated in dollars, such as the cost of bottled water. Intangible benefits, such as improved growing conditions for lawns and plants, cannot be readily expressed in monetary terms.

The revised explanation is longer, but now the path is more clearly laid out. Each paragraph break provides a breathing spell, offering the reader a moment to regroup.

How long should a paragraph be? Once again, a writer must analyze the subject matter. The discussion of economic benefits seemed sufficiently complex to warrant three paragraphs. Now, contrast that discussion with the following basic explanation of words:

For this discussion, let's classify words as simple or complex, and abstract or concrete. Simple words, such as *color* and *native,* communicate immediately, whereas their more complex synonyms, *pigmentation* and *indigenous,* may add to a reader's time and effort. Abstract words, such as *flower* and *animal,* communicate imprecisely because readers will interpret them differently. Concrete words, such as *climbing red rose* and *brown cocker spaniel,* will convey identical word pictures to almost every reader.

In that paragraph, the writer also introduces four separate ideas—

simple, complex, abstract, and concrete. Yet the ideas and definitions are so elementary that few readers would be confused. The reason for the difference in comprehension time between the paragraph on benefits and that on words should be obvious: most of us are unfamiliar with the language of economics, whereas virtually all of us deal with words in our daily lives.

Bridge the Gap with Transitional Words and Phrases

Frequently, and often without realizing it, writers fail to show the connection between ideas in adjoining sentences. When you intend a sentence as an extension of, or a contrast to, an immediately preceding sentence, you should make the connection clear. Otherwise, readers may be left wondering what you are trying to establish. Some may miss the connection altogether.

Provide Transition Within Paragraphs

In the following two sentences, the link between contrasting ideas is missing:

As we use water, portions of it evaporate, transpire, or are lost to the ocean. Significant quantities are returned to surface- and ground-water systems and are reused.

One bridge word will clarify the contrast:

As we use water, portions of it evaporate, transpire, or are lost to the ocean. *Yet,* significant quantities are returned to surface- and ground-water systems and are reused.

In the next example the writer forgot an important bridge word:

To effectively reduce water use we must show consumers how they can save money by using less water. The increase in water costs for increasing quantities used must be large enough to encourage water conservation.

And now the missing link:

To effectively reduce water use we must show consumers how they can save money by using less water. *Therefore,* the increase in water costs for increasing quantities used must be large enough to encourage water conservation.

Next, an entire paragraph with the links missing:

> The reclamation and reuse of waste water present a potential source of additional water supply. Perhaps 60 percent of the treated waste water now discharged to the ocean could be reused. The increased use of reclaimed water would not eliminate the need for additional supplies of fresh water. In most communities, about 50 percent of the water supply is consumed and is unavailable for reclamation.

Now, let's bridge the gap between the ideas in that paragraph:

> The reclamation and reuse of waste water present a potential source of additional water supply. Perhaps 60 percent of the treated waste water now discharged to the ocean could *instead* be reused. *On the other hand,* the increased use of reclaimed water would not eliminate the need for additional fresh water *altogether.* In most communities, about 50 percent of the water supply is consumed and *therefore* is unavailable for reclamation.

In all of the brief paragraphs above, note how the addition of an extra word or two helps clarify the writer's points. The following words and phrases are always available to help writers connect related ideas:

1. *Simple conjunctions*

and	but	for	whereas

2. *Adverbs and conjunctive adverbs**

also	first	finally	ultimately
eventually	lastly	frequently	besides
however	often	moreover	consequently
hence	furthermore	meanwhile	nevertheless
incidentally	likewise	therefore	still

3. *Brief phrases*

in brief	to begin with	of course	in addition
in contrast	stated simply	on the other hand	in essence
in one sense	in other words	more specifically	in fact
in general	what's more	for that matter	by far

Remember, it is your job as the writer to provide transition between closely related ideas.

***Conjunctive adverbs** are adverbs used as connectives (**conjunctions**), usually between sentences or between **clauses** separated with a semicolon. See also **Adverbs; Semicolon** in the Handbook.

Provide Transition Between Paragraphs

You may often write a paragraph as an extension of, or a contrast to, a preceding paragraph. Once again, you must be careful to bridge the gap, or readers may miss the connection altogether. In the two paragraphs that follow, the link between contrasting ideas is missing:

> The establishment of new communities in previously unoccupied areas would present opportunities to use new techniques of waste disposal. Through careful planning, we could design new, more efficient disposal facilities that might never become inadequate.
>
> The mere construction of new waste-disposal facilities would not in itself end the threat of eventual waste overloads. Equally important would be carefully designed growth and zoning controls.

As you may have discovered, the second paragraph lacks one brief phrase that would quickly identify the contrasting ideas. Look at them again, this time with the "bridge" restored.

> *On the other hand,* the mere construction of new waste-disposal facilities . . .

Finally, in these two paragraphs that discuss sewage disposal to streams, the links between related ideas within each paragraph, and between the two paragraphs themselves, are missing:

> Many communities once discharged their wastes to streams and rivers without great regard for the consequences. Bacteria and other microorganisms converted the sewage and other organic matter into new bacterial cells, carbon dioxide, and other products. Bacteria cannot perform this chore without oxygen—oxygen the stream acquires from the air and from plants growing in the water. This dissolved oxygen in the stream is essential to fish and other aquatic life.
>
> If only nominal amounts of sewage are discharged to a stream, the oxygen used to neutralize it is quickly restored, and aquatic life is not significantly affected. When sewage loads become excessive, taking more dissolved oxygen from the water than can readily be restored, the sewage will decay and the water will begin to give off unpleasant odors. Unless the waste loads are reduced, the stream will eventually lose all of its oxygen.

Now, let's connect the related thoughts in the two paragraphs and see how the writer's ideas fairly leap off the page at the reader:

> *Only a few years ago,* many communities discharged their wastes to rivers and streams without great regard for the consequences. Bacteria and other microorganisms converted the sewage and other organic matter into new

bacterial cells, carbon dioxide, and other products. *However,* bacteria cannot perform this chore without oxygen—oxygen the stream acquires from plants that grow in the water *itself.* The dissolved oxygen in the stream is also essential to fish and other aquatic organisms; *without it, they cannot survive.*

Of course, if only nominal amounts of sewage are discharged to a stream, the oxygen used to neutralize it is quickly restored, and aquatic life is not significantly affected. When sewage loads become excessive, *however,* taking more dissolved oxygen from the water than can readily be restored, the sewage will decay and the water will soon begin to give off unpleasant odors. Unless the waste loads are reduced, the stream will eventually lose all of its oxygen, *and the water will literally die.*

Note how, in the second version, the addition of a few transitional words and phrases helps unify the entire discussion. Remember, it is your job to bridge the gap between related ideas. Remember, too, that scores of transitional words and phrases—such as those on page 30—are available to help light the way for readers.

Revise, Revise, Revise

Revision is a crucial, and often neglected, step in the preparation of any writing. Countless times over the past few years, I have decorated the margins of hastily prepared papers with question marks and brief instructions such as, "Say again, please." Most, I have long suspected, were products of the midnight oil, written such a short time before that I could all but smell the fresh ink.

Revision is your opportunity to seek out the sagging syntax, the grammatical goof, and the dangling modifier. Never will I forget this classic introduction to an imaginative story about flying saucers:

Hurtling through the night sky, emitting showers of sparks and loud whooshing sounds, the wide-eyed boys watched the silvery object in startled wonderment.

Then there was this description of a hydroelectric project:

Hydroelectric power is produced by turbines at the base of a dam turned by the falling water.

Question: Was the dam on a rotating base?

By carefully reading a first draft, you can usually avoid errors like those. Therefore, whenever an idea is fermenting within you, first take time to organize it. Then write a first draft as fast as you can get the words on paper, without regard for perfect grammar, punctuation, spelling, and the other proprieties. Neither pause to look up words nor overly concern yourself with the other blunders that invariably turn up in a first effort. In short, don't break your concentration or shut off the creative juices that are inspiring you at the moment. When you have finished the first draft, walk away and stay away as long as possible, at least until you can face it with a critical eye.

Now Proofread Carefully

When you return, your first effort will probably look different to you, and you can begin surgery—the search for misspelled words, malapropisms, unbalanced sentences, errant grammar and punctuation, and the "things you didn't really mean." To locate spelling errors, read each sentence backward (from right to left); to find the other slips just mentioned, read the paper line by line and sentence by sentence. Use a dictionary and the Handbook as you revise.

Did You Say What You Meant to Say?

Revising can be more than merely correcting errors. As you read your first effort carefully, you may decide that some of it is repetitive or irrelevant, or that it needs additional material. You may also want to replace some words and phrases with more exact words and phrases. You may want to rewrite the entire paper to give it stronger focus. In short, taking time to revise, at least once, everything you write will enable you to say what you really want to say.

Professional writers may revise a piece two, three, four, and more times before both writer and editor are satisfied. Surely you can find time for at least one revision and rewriting. Instead of writing and praying, follow these five steps:

1. Organize
2. Write
3. Relax
4. Revise
5. Rewrite

PART TWO

A Technical Writer's Handbook

A

A, AN. *A* is the indefinite article, which is used to modify nouns or noun phrases that denote a single but unspecified item or person. The general rule is to use *a* before a word beginning with a consonant and *an* when a word begins with a vowel. Regardless of the spelling, however, when the modified word begins with a consonant sound, the correct article is *a*. Some examples:

1. I'll be there in *an* hour (vowel sound—our).
2. This is *a* historic moment (consonant sound—hiss).
3. He bought *a* uniform for the parade (consonant sound—yu). BUT—
4. You had better carry *an* umbrella (vowel sound—um).

ABBREVIATIONS. Abbreviations are usually acceptable in technical reports and instruction manuals, particularly when technical terms are used repeatedly. In much other exposition, and particularly in articles intended for wide distribution, abbreviations should be used sparingly. (Magazines, newspapers, and many business organizations set their own standards for abbreviations; most follow established conventions.)

The following generally accepted conventions are applicable to most technical writing:

1. The expressions *a.m., p.m., A.D.,* and *B.C.* are acceptable in almost all writing.
2. Abbreviations for such titles as Ph.D., C.P.A., and M.S. are used following names. Standing alone, however, such terms should be written out, normally in small letters (although academic degrees are usually capitalized when the full name of the degree is used). For example:

a. Edgar Thompson, B.S.E.E. (BUT) He is an electrical engineer.
b. John Smith, C.P.A. (BUT) He is a certified public accountant.
c. Linda Roberts, L.L.B. (BUT) She is an attorney.
d. Mary Jones, M.A. (BUT) She holds a Master of Arts degree.

3. Such titles as *Professor, President, Representative, Manager, Governor* should always be written out. Similarly, terms of respect, such as *Reverend* and *Honorable* (The *Honorable* James J. Brown, for example) are also written out and should never be abbreviated (not the *Hon.* James J. Brown).
4. Never abbreviate a person's name—Wm., Jas., Robt., etc.
5. Recently, abbreviations of certain terms, particularly names of government agencies and organizations, have become so well known that they are often expressed as abbreviations, for example, FBI, CIA, IRS. Unless you suspect that some readers may not understand them, you may abbreviate most well-known "spoken" abbreviations in all but the most formal writing. See also **Acronym**.
6. The abbreviations *e.g.* (for *exempli gratia,* Latin for *for example*) and *i.e.* (*id est,* Latin for *that is*) may be used in most situations. However, they should be used correctly. Writers often interchange the two terms and thus confuse the issue. Used excessively, moreover, the terms become wearisome, and technical writers might better use the English equivalents.
7. Abbreviations are normally capitalized only when the terms for which they stand would be capitalized. However, **acronyms** and certain other abbreviations are capitalized to (a) distinguish them from ordinary words with which they could be confused, or (b) emphasize that they are abbreviations. An example of (a): LED (light-emitting diode); of (b) MSF (multistage flash). For examples of how such abbreviations are used in text, see item 8 below.
8. You may sometimes have to use certain terms, particularly units of measure, over and over again. In such cases you may prefer to abbreviate them, if only to prevent the repetitious terminology from dominating the discussion and overwhelming readers. In technical writing, such terms as *cubic feet per second, revolutions per minute,* and *pounds per square inch absolute* may appear several times in a single paragraph and sometimes in a single sentence. In such instances, readers will probably appreciate the use of abbreviations, which will cut down on the repetitious terminology.

   If you are concerned that some readers may not understand the abbreviations, you can follow the conventional practice of writing out the term the first time you use it, then giving the abbreviation in parentheses. From then on, simply use the abbreviation.

Here are two examples of abbreviations used to suppress repetitious terminology:

a. "On March 3, peak flood flows were recorded at 91,000 cubic feet per second (cfs). By March 5, however, the flows had receded to 64,000 cfs. On March 8, flows increased slightly to 67,000 cfs but by March 15, were down to 31,000 cfs, normal for this time of year."

b. "The capacity of the prototype desalting plant is 40 million gallons per day (mgd). Two 20-mgd units were selected for the 40-mgd plant. This represents both a step-up in technology and a significant increase in unit size. The requirement that the steam be obtained from two power units also lends itself to the division of the 40-mgd plant into two 20-mgd units.

"The multistage flash (MSF) distillation process will be used in the prototype plant. The MSF module at the San Diego test facility has been used successfully since 1970. The facility was designed for a large-scale plant of up to 40-mgd capacity. The MSF process is the most advanced distillation process and has undergone extensive testing.

"Some of the largest MSF plants now operating, along with their capacities, are shown in table 4."

9. One form of abbreviation writers should treat with care is the use of symbols for words. You will serve your readers best by observing these conventions:
   a. Use no symbols (in text) except the degree sign (°) and the dollar sign ($).
   b. Write out *percent* (instead of %), and don't use the symbols ′ for feet or ″ for inches. These last two are especially risky, because readers can easily misinterpret them. However, the symbols ′ and ″ are acceptable in drawings and blueprints; the symbols for degrees (°), minutes (′), and seconds (″) are used in expressions of latitude and longitude, 41° 15′ 20″ N, for example.
   c. One final practice to avoid is the use of a dash or hyphen in place of the word *to*. For example, write, "The depths varied from 20 *to* 30 feet (not 20–30 feet)."
10. Three other conventions you should be aware of:

a. Units of measure are abbreviated only when they follow a figure; for example, 65 ft, 450 rpm. However, such terms are not abbreviated when expressed without a number, as in the sentence, "Convert the dimensions from feet to inches."
b. Abbreviations of units of measure are expressed in singular form: 36 lb, not 36 lbs.
c. Unless they would otherwise spell a word, abbreviations of units of measure need no periods. For example, 14 in. (to avoid confusion with the word *in*); but 75 ft (no period necessary).

The myriad of abbreviations used in different disciplines makes it virtually impossible to present an all-inclusive list. There are abbreviations for military titles, for units of measure, for physics and chemistry, and for commerce and industry, to name only a few. The aerospace industry has a great many abbreviations that would be irrelevant to most other fields. In addition, there are a host of mathematical symbols, including a number of Greek letters with arbitrary meanings. Then there are the commonly used abbreviations in **footnotes,** chiefly Latin, most of which date back to the Middle Ages but which are still found in scholarly books and papers. The most common of those are listed on pages 116–17.

Technical writers will be concerned most often with the abbreviations used in science and industry. A word of caution: there is no all-inclusive industrywide list of abbreviations. Although the American Society of Mechanical Engineers (ASME) does publish a list of standard letter symbols for use in science and technology, the abbreviations actually used in technical writing vary from industry to industry and among different disciplines. For example, the ASME abbreviation for *cubic feet per second* is ft^3/s; yet, many organizations use *cfs.*

Therefore, writers who use abbreviations are advised to heed paragraphs 8, 9, and 10 above. Whether to abbreviate is a question demanding a writer's judgment. Some general advice: when in doubt, don't abbreviate.

In this book:

1. See **Documentation** for abbreviations used in **footnotes** (pp. 116–17).
2. See **Correspondence** for the state two-letter address codes (p. 92).
3. See **Proofreading** for a list of proofreader's marks (p. 205).
4. See **Metric System** for a number of abbreviations and symbols applicable to the metric system (p. 168).

Here is a generally acceptable list of abbreviations often used in scientific and technical writing; following the alphabetical list is a list of mathematical symbols. As we discussed earlier, neither list is intended to be all-inclusive.

Common Abbreviations

absolute	abs	continuous wave	cw
air horsepower	ahp	cosine	cos
alternating current	ac	cotangent	cot
(adjectival form)	a-c	cubic centimeter	cc or cm^3
ampere	amp		
ampere-hour	amp-hr	cubic feet	cu ft or ft^3
amplitude modulation	AM	cubic feet per second	cfs
Angstrom unit	A	cubic inch	cu in. or in.^{3}
atmosphere	atm		
atomic weight	at wt	cubic meter	cu m or m^3
audio frequency	AF		
avoirdupois	avdp	cubic yard	cu yd or yd^3
barometer	bar.	current	I
barrels	bbl	cycles per second	cps
billion electron volts	bev	decibels	dB
biochemical oxygen demand	bod	decigrams	dg
board-feet	bd-ft	deciliters	dl
brake horsepower	bh	decimeters	dm
Brinell hardness number	Bhn	degree	deg or °
		dekagrams	dkg
British thermal units	Btu	dekaliters	dkl
calorie	cal	dekameters	dkm
Celsius (or Centigrade)	C	dewpoint	dp
		diameter	dia
center of gravity	cg	direct current	dc
centimeter	cm	(adjectival form)	d-c
circumference	circum	dozen	doz
cologarithm	colog	drams	dr

electromotive force	emf
electron volts	ev
elevation	el
equivalent	equiv
Fahrenheit	F
farads	f
feet	ft
square feet	ft^2
cubic feet	ft^3
feet per second	ft/sec
foot-candle	ft-c
foot-pound	ft-lb
frequency modulation	FM
gallons	gal
gallons per day	gpd
gallons per minute	gpm
grains	gr
grams	gm
gravitational acceleration (32 ft/sec^2)	g
hectare	ha
hectoliter	hl
hectometer	hm
square hectometer	hm^2
cubic hectometer	hm^3
henries	h
high frequency	hf
horsepower	hp
horsepower-hours	hp-hr
hours	hr
hundredweight	cwt
inch	in.
square inches	$in.^2$
cubic inches	$in.^3$
inch-pounds	in.-lb
infrared	IF
inside diameter	ID
intermediate frequency	i.f.
joules	j
Kelvin (temperature)	K
kilocalories	kcal
kilocycles	kc
kilocycles per second	kc/sec
kilograms	kg
kilojoules	kj
kiloliters	kl
kilometers	km
square kilometers	km^2
cubic kilometers	km^3
kilovolts	kv
kilovolt-amperes	kva
kilowatt-hours	kw-hr
kilowatts	kw
lambert	L
latitude	lat
length	l
linear	lin
linear foot	lin ft
liter	l
logarithm	log.

longitude	long.
low frequency	l.f.
lumen	lm
lumen-hour	lm-hr
megacycles	mc
megawatts	mw
meters	m
square meters	m^2
cubic meters	m^3
microamperes	μamp*
microinches	μin.
microseconds	μsec
miles per gallon	mpg
miles per hour	mph
milliamperes	ma
millibars	mb
millifarads	mf
milligrams	mg
milliliters	ml
millimeters	mm
millivolts	mv
milliwatts	mw
minutes	min
nautical miles	NM
negative (polarity)	neg or −
number	no.
octane	oct
ounces	oz
outside diameter	OD
parts per billion	ppb
parts per million	ppm
positive (polarity)	pos or +
pounds	lb
pounds per square inch absolute	psia
pounds per square inch gauge	psig
radio frequency	RF
radium	r
Rankine (temperature)	R
resistance (electrical)	R
revolutions per minute	rpm
roentgens (radiation)	r
secant	sec
seconds	sec
specific gravity	sp gr
square foot	ft^2
square inch	$in.^2$
square meter	m^2
square mile	mi^2
tablespoonful	tbsp
tangent	tan.
teaspoonful	tsp
temperature	temp
tensile strength	ts
thousand	M
tons	t
ultra high frequency	UHF
vacuum	vac

*See also **Metric System** for use of the symbol μ (mu).

very high frequency	VHF	watts	w
volt-amperes	va	wavelength	wl
volts	v	weight	wt
volume	vol	yards	yd
watt-hours	whr	years	yr

Mathematical Symbols

$\pm$	plus or minus	$\div\!\!:$	geometrical proportion
$\times$ or $\cdot$	multiplied by	$\therefore$	therefore
$\div$	divided by	$\because$	because
$<$	is less than	$'$	minute
$>$	is greater than	$''$	second
$\nless$	not less than	$^\circ$	degree
$\ngtr$	not greater than	$\parallel$	parallel
$\leq$	is equal to or less than	$\nparallel$	not parallel
$\geq$	is equal to or greater than	$\int$	integral
$\equiv$	is identical to	∞	infinity
$\neq$	is not equal to	$\sqrt{\ }$	square root
$\approx$	is approximately equal to	$\sqrt[3]{\ }$ $\sqrt[4]{\ }$	cube root, fourth root, etc.
$\sim$	is similar to	$'$ $''$	single prime, double prime, etc.
$\underline{\underline{m}}$	is measured by		
$:$	ratio	$\rightarrow$	forms, yields
$\leftarrow$	is formed by	$\sum$	summation of
$\rightleftarrows$	forms and is formed by	π	pi (3.1416)
∂	partial differential	ϵ or e	base (2.718) of natural system of logarithms
$::$	proportion		

ABOUT, AROUND. *About* and *around* should not be confused. Strictly speaking, *around* means starting at one point and returning to that point. Colloquially, *around* is often used in place of *about* (approximately), as in the sentence, "The meeting will start *around* 2:30." In precise writing, however, the two words should not be interchanged.

ABOVE, THE ABOVE, THE AFOREMENTIONED, THE AFORESAID. These adjectives have long been used as vague references to something mentioned previously. For instance, "the *above mentioned,* or the *aforesaid* (or *aforementioned*) lakes." Since the construction is unnatural and pompous, replace it with a definite reference, such as, "the lakes discussed on page 10" or "the lakes mentioned in the preceding paragraph."

ABSOLUTE PHRASE. An absolute phrase is a modifier containing a minimum of functional words (**prepositions** and **adverbs**). It functions like an **adverbial clause** or a **prepositional phrase**. For instance, you might write, "When the play ended (adverbial clause), we went home." (OR) "After the play (prepositional phrase) was over, we went home." In the absolute construction, the statement might be, "The play over (absolute phrase), we went home."

When the absolute phrase contains a participle, it is called the *ablative absolute* (from the Latin absolute construction). Note how it is used in these sentences:

Adverbial clause: When the stress tests were completed, the technician averaged the results.
Ablative absolute: The stress tests completed, the technician averaged the results.

Prepositional phrase: The fire engine roared down Main Street *with its red lights blazing and its siren shrieking.*
Ablative absolute: Its red lights blazing and its siren shrieking, the fire engine roared down Main Street.

As you can see, the absolute phrase (modifier) puts across the full meaning of a **clause** or **prepositional phrase** with a minimum of words. Its strength lies in its conciseness, enabling a writer to produce an effective **modifier**—frequently one that is stronger than a full-blown clause or phrase—with no wasted words.

ABSTRACT WORDS. See *Abstract and Concrete Words* in part 1, p. 15.

ACCEPT, EXCEPT. To *accept* (verb) is to receive, admit, or take on a responsibility. *Except* is usually a **preposition** meaning *other than* or a **conjunction** meaning *if it weren't for the fact that.* For instance:

1. John *accepted* the award for the entire laboratory staff.

2. When he took that job, he *accepted* full responsibility for finishing the work on time.
3. Everyone *except* (preposition) Jane went to the meeting.
4. I would buy that computer in a minute *except* (conjunction) that it costs $120,000.

ACRONYM. An acronym is a word formed with the initial letters of a title or name, such as UNESCO (United Nations Educational, Scientific and Cultural Organization) or LED (light-emitting diode). Like certain spoken abbreviations (FBI, IRS, UFO, etc.), a number of acronyms have now become household words and ordinarily may be used in technical writing and other exposition.

As with **abbreviations,** if you are concerned that some readers may not understand the acronym, write out the term the first time you use it, following it with the acronym in parentheses. That is probably the safest procedure with all but the most obvious terms.

A few acronyms have become so commonplace that many readers will not know that the words were originally derived from longer forms. For example, *laser* (light amplification by stimulated emission of radiation); *radar* (radio detection and ranging); NERVA (nuclear engine for rocket vehicle application); OPEC (Organization of Petroleum Exporting Countries).

Note that acronyms are written without periods. See also **Abbreviations**.

ACTIVATE, ACTUATE. Although both words mean "to make active," *activate* is properly used with physical or chemical processes, whereas *actuate* denotes a mechanical process. For example:

1. *Activated* carbon can be used as a filter to remove tastes and odors from water.
2. The flight-control system *actuates* the jetevators, which move into and out of the exhaust stream and stabilize the rocket during powered flight.

ACTIVE VOICE. See **Voice**.

ADAPT, ADOPT. To *adapt* a plan is to modify it for another use. To *adopt* a plan is to take it over and put it into use. For instance:

1. The proposed test plan was *adopted* by the laboratory.
2. The laboratory test plan was *adapted* for use by the manufacturing division.

ADDRESSES. See **Correspondence.**

ADJECTIVE.

1. Adjectives modify—and thus describe or limit the meaning of—**nouns** and **pronouns:** *green* car, *heavy* book, *poor* me, *lucky* you.
2. Other parts of speech may function as adjectives:
   Nouns: floor joist, *freeway* traffic, *stress* test
   Verbs: drinking water, *running* track, *well-deserved* reward
   (This form of a verb is called a **participle;** when a participle serves as an adjective, it is called a verbal adjective.)
3. Adjectives may also be classified as *common* or *proper.*
   Common: heavy book, *wide* street
   Proper: Spanish architecture, *Roman* arch
4. One other adjectival form is discussed under **Preposition.**

See also **Comparative and Superlative of Adjectives and Adverbs.**

ADVERB.

1. Adverbs modify **adjectives,** other adverbs, **verbs,** and entire **clauses:**

   With an adjective: The situation is *highly* complex.

   With another adverb: The roof was *very carefully* inspected.

   With a verb: He wrote *hurriedly.*

   With a clause: Perhaps it will rain tonight.

2. Adverbs are often formed with **adjectives** or **participles** plus *ly:* barely, closely, deeply, firmly, independently, slowly, wrongly, amazingly, decidedly, deservedly, markedly, exceedingly.
3. Some adverbs have the same form as adjectives: *bad* (I feel bad); *better* (You may feel better tomorrow); *good* (I feel good today); *loud* (He talks too loud); *slow* (Drive slow through town); *tight* (Hold on tight now). Others include close, early, even, fair, fast, first, late, much, right, smooth.
4. Adverbs are often classified by meaning; that is, adverbs of:

   degree (to what extent): almost, altogether, nearly, entirely, exceedingly, enormously, principally, somewhat

   manner (how): awkwardly, boastfully, carefully, graciously, hurriedly, naturally, rashly

place (where): below, forward, here, there, thereabouts
time (when): immediately, lately, soon, when

5. Another type of adverb is the *conjunctive* adverb, which is often used as a connective between clauses and sentences. Some of the most common conjunctive adverbs are *accordingly, also, consequently, furthermore, however, moreover, nevertheless, therefore, yet.* When conjunctive adverbs are used between independent clauses, the clauses are not separated with a comma; instead, a semicolon is used:
   a. We ran out of money; therefore, we discontinued the test program.
   b. The test results were disappointing; however, they came as no surprise.
6. One other adverbial function is discussed under **Prepositions.**

See also **Comparative and Superlative of Adjectives and Adverbs.**

ADVERBIAL CLAUSE. An adverbial clause is a subordinate clause introduced by an adverb; adverbial clauses modify verbs, adjectives, or adverbs in the main clause of a sentence. For instance:

1. I enlisted in Chicago, *where I had lived for twenty-one years* (modifies the adverbial **prepositional phrase** *in Chicago*).
2. The raft overturned *where the current was swiftest* (modifies the verb *overturned*).
3. The air was cold *when we went outside* (modifies the adjective *cold*).

ADVICE, ADVISE. *Advice* is the noun; *advise* is a verb. When you *advise* someone, you offer *advice.*

AFFECT, EFFECT. To distinguish between these two troublemakers, remember that *affect* is almost always a verb. (Its only use as a noun is restricted to psychology.) *Affect* is also used in the sense of *pretend* or *feign. Effect* is usually a noun, although it can also be a verb (meaning *to put into effect*). Some examples:

1. The city was not *affected* by the flood. (The flood had no *effect* on it.)
2. She *affected* a liking for (pretended to like) calculus.
3. The *effects* of the flood were widespread.
4. The housing program was *effected* (put into effect) with minimum delay.

AFFINITY, APTITUDE. *Affinity* denotes a natural attraction between chemical elements or, in another sense, between two persons. It does not mean *aptitude,* however. For instance:

Wrong: He seems to have a natural *affinity* for science.
Right: He seems to have an *aptitude* for science.
Right: Hemoglobin has a greater *affinity* for carbon monoxide than for oxygen.

AFOREMENTIONED, AFORESAID. See **Above, The Above, The Aforementioned, The Aforesaid.**

AGGRAVATE, ANNOY. Strictly speaking, *aggravate* means to make worse, although it is widely used informally to mean *annoy,* and in conversation that meaning is acceptable. Technical writers, however, should use the word to denote its precise meaning, *worsen.* For example: "The low water level in the reservoir was *aggravated* by six months of subnormal precipitation."

AGREEMENT OF SUBJECT AND VERB; OF PRONOUN AND ANTECEDENT. See **Sentence Types and Construction.**

AGREE ON, AGREE TO, AGREE WITH. To *agree on* something is to be in accord with others; to *agree to* something is to consent; to *agree with* something is to be in accord with it. For instance:

1. The city council *agreed on* a new parking plan.
2. The council *agreed to* raise parking fees.
3. The mayor *agreed with* their decision.

In the last sense, avoid the inflated idiom "agree with the idea that." For instance, "I *agree with the idea that* we need a new test plan" means, "I *agree that* we need a new test plan."

ALL (as a prefix). See **Prefixes and Suffixes.**

ALLERGENIC, ALLERGIC. To be *allergic* is to be sensitive to something, as to chocolate or pollen. The substance causing the allergy is *allergenic.*

ALL READY, ALREADY. Beware of this pair, the first of which simply denotes *preparedness. Already* is an adverb meaning *previously.* For example: "The group was *all ready* (prepared) to leave, but the bus had *already* (previously) departed."

ALL RIGHT, ALRIGHT. Most authorities consider the single-word form *(alright)* substandard. Simply remember that *alright* is not *all right.*

ALL TOGETHER, ALTOGETHER. *All together* denotes a group (together in one place). *Altogether* is an **adverb** meaning *completely* or *entirely.*

ALLUDE, ELUDE. To *allude* is to refer to casually or indirectly. To *elude* is to escape by dexterity, as to elude pursuit. *Elude* also means to escape the mind. For instance: "He *alluded* to the theory of relativity, but his meaning *eluded* me."

ALLUSION, DELUSION, ILLUSION. Distinguish carefully between the members of this trio; the meaning of *allusion* is quite different from the meanings of the other two. An *allusion* is an indirect reference to a similar idea or event; a *delusion* is false hope or erroneous belief (as a delusion that the United States will never run out of natural gas); an *illusion* is an erroneous physical perception like the mirage in the desert that leads the traveler to believe that water lies ahead. Don't be misled by the similar appearance of the three words, as was the student who wrote, "The following *illusion* to Newton's *Principia* will illustrate my point."

ALONG THE LINES OF. This swollen double prepositional phrase means simply *like* or *similar to:*

Inflated: He is developing a new program *along the lines of* last year's safety campaign.
Concise: He is developing a new program *similar to* last year's safety campaign.

A LOT, A LOT OF. *A lot* and *a lot of* are colloquial (conversational) substitutes for *considerable, much, many, a large amount of,* etc. In technical writing, the more formal words would be more appropriate. For example:

Colloquial: We had better give their proposition *a lot of* thought.
Improved: We had better give their proposition *considerable* thought.

Colloquial: They have *a lot of* work to finish before Friday.
Improved: They have *a great amount of* work to finish before Friday.

Colloquial: A lot more study will be required before the problem is solved.
Improved: Solving the problem will require *much* more study.

Colloquial: With today's gasoline prices, *a lot* more people are riding the busses.
Improved: With today's gasoline prices, *many* more people are riding the busses.

See also **Colloquial Language.**

ALTAR, ALTER. The *altar* is in the church; to *alter* a design is to change or modify it.

ALTERNATE, ALTERNATIVE. As an adjective, *alternate* means occurring by turns or every other (as in *alternate* lines); it also means one or the other (but not more than two). Its only correct use as a noun is as one person designated to take the place of another, for instance, at a convention.

Strictly speaking, *alternative* means the *second of two.* Modern usage, however, has broadened its meaning to include one of a number of items or courses offered as choices. As an adjective, *alternative* means offering a choice of two or more things. Some examples:

1. Because the chairman was ill, he sent an *alternate.*
2. That procedure must be followed; the *alternative* is chaos.
3. An *alternative* choice; a third *alternative.*

AMBIGUITY. Ambiguity, in which there is a possibility of confusing two meanings, often results from **misplaced modifiers** and from the careless use of **pronouns.** For a discussion of modifiers, see *Be Careful with Modifiers* in part 1, p. 20. Reference of pronouns is discussed in **Sentence Types and Construction.** Ambiguity can also arise from the careless use of words. See **Homonyms; Malapropisms.**

AMONG, BETWEEN. The general rule is that *between* is used with two persons or things; *among* is used with three or more. For instance:

1. The designer had to choose *between* steel and aluminum.
2. The storage space was divided *among* the four departments.

However, *between* is also used with more than two to distinguish between any two of them. For example: "Statisticians distinguish carefully *between* the meanings of *median, mean,* and *average.*"

AMOUNT, NUMBER. *Amount* is used with something viewed in bulk; *number* is used with items that can be counted. For example, an *amount* of salt, but a *number* of test tubes.

ANALYSIS, ORGANIZATION OF. The purpose of an analysis is to examine something—to break it down into parts to learn the facts about it. Just as a chemist analyzes a sample of water by determining the concentrations of the various minerals it contains, a corporate controller might analyze a steep rise in business expenses to determine why they have suddenly increased. If the analyst can determine the cause, he or she can then take steps to rectify or mitigate the problem.

Analysis requires digging beneath the surface, so to speak, to determine actual facts. It usually requires that the analyst answer a number of questions and frequently reveals facts that differ from popular beliefs and rumors. For instance, if armed robberies were increasing in a particular city, the residents might begin to believe that the police were inefficient or understaffed. Yet, an analysis of where and when the robberies were occurring might reveal that perhaps the police were merely in the wrong parts of the city at certain times.

The analysis would require answers to such questions as:

1. What types of crime are increasing?
2. At what times are they occurring most frequently?
3. In what parts of thc city arc they increasing?
4. In what parts of the city are there no crimes at all?
5. How long does it take the police to respond to calls?
6. Would special warning systems help?

With answers to these and other questions, the police department might be able to increase their coverage and surveillance in certain areas, perhaps decreasing the number of robberies without increasing the number of officers on the force.

That example is necessarily oversimplified—there would be many more questions than the six just listed. It does, however, give some idea of how analysis is used. And once a writer has the facts—the underlying cause of a problem—writing an analysis is not difficult. Often, a writer will develop an analysis with a cause-and-effect pattern (for these reasons, that is happening).

To cite an actual example, here is part of an analysis prepared by an engineering consultant of the excessively hard water supplied to a south-western community. The consultant was hired to determine whether installation of a central water-treatment facility would be economical.

The hard municipal water causes numerous problems for all residents. For instance, washing machines and dishwashers require frequent repair because of sticking solenoid valves. Water heaters must be replaced every five years, whereas with softer water they are normally guaranteed for ten years. The hard water also stains and discolors plumbing fixtures, faucets, and mirrors; shower heads are continually plugged.

Very few persons use the community water supplies for cooking and drinking, but instead purchase bottled or "bulk" water delivered by truck. Those who use bulk water must have a 100-gallon tank, which is connected to a third faucet in the kitchen. The delivered cost of bulk water is $0.13 per gallon; the average monthly cost per household is $14. The delivered cost of bottled water is $1.61 for five gallons. Average weekly consumption of bottled water ranges from seven gallons per person during the winter to ten gallons per person during the summer.

Water softeners are used in about 10 percent of the homes. Those who have them report a 50 percent savings in the cost of soap and washing powders. The cost of owning and maintaining a water softener averages about $10 per month.

AND ETC. Because *etc.* (Latin *et cetera;* English *and so forth*) contains an *and,* the use of *and etc.* is redundant. Simply use *etc.* alone. See also **Etc.**

AND/OR. This expression, particularly cherished by governmental and technical writers, really indicates a choice among three courses or things. A and/or B, for instance, equals: (1) A, or (2) B, or (3) A and B. Why not omit the confusing locution and say what you mean? For instance, "Solid wastes can be disposed of by burning and/or burying" means: "Solid wastes can be disposed of by burning, by burying, or by both methods."

ANOTHER, ADDITIONAL. Strictly speaking, *another* means *one more* and should not be used loosely to denote more than one; in the latter case, the correct adjective is *additional* or *more.* For example:

Loose: Another ten engineers will be hired next month.
Precise: An *additional* ten engineers will be hired next month. (OR) Ten more engineers will be hired next month.

ANTECEDENT. An antecedent is the word to which a **pronoun** refers. The need for careful use of pronouns and their relationships to antecedents are discussed under **Sentence Types and Construction.**

ANTI (as a prefix). See **Prefixes and Suffixes.**

ANTICIPATORY SUBJECT. See **Expletive.**

ANXIOUS, EAGER. To be *anxious* is to have feelings of anxiety about a forthcoming event. To be *eager* is to feel strong interest or desire. For example:

1. He is *anxiously* (with some anxiety) looking forward to the examination, the outcome of which will determine his eligibility for promotion.
2. Every morning she drove to work *eagerly* (with desire), for she truly enjoyed her job as a programmer.

APOSTROPHE. The apostrophe causes problems for some writers, although its functions are not difficult to master. Use it:

1. To indicate possession:

   a. *Helen's* book (singular possessive—apostrophe and *s*)

   b. the *workers'* tools (plural possessive—apostrophe only)

   Note: With plural words that do not end in *s,* the *s* is added; for example, the *men's* wages. With singular words that end in *s*—particularly names—the *s* is also added, unless the resulting possessive form would be difficult to pronounce. For example, Mr. *Jones's* house but Henry *Peters'* automobile.

   Opinions differ on the use of an apostrophe to indicate possession by inanimate objects: the *automobile's* engine, the *factory's* smokestacks, the *book's* cover. Actually, the use of the possessive in such constructions is usually unnecessary: *automobile* engine, *factory* smokestacks, *book* cover. Somehow, the use of the noun modifier instead of the possessive seems more graceful and natural. See also **Case.**
2. With *s* to form the possessive of inanimate objects denoting time, measure, or space:

Singular	*Plural*
a day's work	two days' pay
a snail's pace	two weeks' notice
at arm's length	five dollars' worth

3. To indicate a contraction: *can't, you're, it's, I'll. Caution:* Don't confuse *it's* (meaning *it is*)with *its* (singular possessive pronoun that requires no apostrophe). The two are often interchanged, as are *your* (possessive) and *you're* (*you are*). See also **Its, It's; Who's, Whose; Your, You're.**

4. To designate the plural form of numbers and individual letters: the 1970's, three c's, five 4's, etc. The trend is to omit the apostrophe in such constructions. Witl a single letter ornumber, however, the apostrophe is needed to prevent confusion:

   a. The population increased more slowly during the *1970s*.

   b. There are two *c's* and two *m's* in *accommodate*.

Caution: Don't use an apostrophe in simple plural nouns: *boy's, book's,* etc. The plural form is *boys, books,* etc.

APPLICATION LETTERS AND RÉSUMÉS. In a letter applying for a job, come straight to the point—perhaps stating how you learned about the opening—and then give your specific qualifications. To avoid overwhelming the reader with details, keep the letter brief and enclose a résumé with complete but *relevant* information.

Natural common mistakes in an application letter are (1) including too much irrelevant information, and (2) writing the letter in terms of what the job can do for the writer, instead of what he or she can do for the prospective employer. Avoid such self-serving statements as, "I have always wanted to be a programmer," "I have always wanted to live in Idaho," or "My wife and I love small towns," which writers often include, believing they will enhance their chances of getting the job.

In the closing of your letter, avoid the often-used "If my background fits your requirements, I will be glad to come for an interview." That seems to suggest that the applicant may not be qualified after all. Therefore, shun the conditional clause, and simply offer to report for an interview at the employer's convenience.

Remember:

1. Employers often receive a great many application letters for a single job, particularly when the position is widely advertised. Therefore, they appreciate brief letters. Save the pertinent details for the résumé.
2. The purpose of your letter is not to get the job through the mail—it is to obtain an interview, at which you can bring up other pertinent points in response to specific questions.
3. This is your first contact with a prospective employer, who will form a first impression of you after reading your letter and résumé. Therefore, make certain that both letter and résumé are neat, attractive, and free from grammatical, punctuation, and spelling errors.
4. What is the employer most concerned with? *Your ability to fill a particular job.*

Here are two sample letters of application and accompanying résumés. The first is from a recent college graduate with limited part-time experience. The second is from an applicant with several years' experience as a television news reporter. Note that the second résumé (from the news reporter) is in seminarrative style, a change from the impersonal "did this, did that" approach.

7824 Entrance Street
Fair Oaks, California 95628

June 10, 1981

Mr. Gerald Chapman, Manager
Master Data Corporation
3600 First Street
Central City, California 90004

Dear Mr. Chapman:

Mr. Charles White of the ABC Records Corporation in Sacramento has informed me that you plan to add a beginning programmer-analyst to your staff. I would like to be considered for that position. I have recently graduated from California State University in Sacramento with a degree in Business Administration. My special area is computer programming and analysis.

Along with my formal education, I have been fortunate enough to obtain some practical experience also. Last summer, I worked full-time as an assistant programmer at ABC Records Corporation, where I was able to apply my knowledge of COBOL to inventory control and payroll. And, during my last year in school, I worked part-time in the Business Office here at California State, where I gained additional programming experience.

I have listed my education and experience, along with three references, on the enclosed résumé. I can report for an interview whenever it is convenient for you.

Very truly yours,

Rita J. Henly

(Miss) Rita J. Henly

Enclosure: 1

Résumé

Rita J. Henly
7824 Entrance Street
Fair Oaks, CA 95628
(916) 961-3441

Personal Data

Age: 24
Health: good
Marital status: single

Education

California State University, Sacramento
Degree: B.S., Business Administration, June 1981
Special emphasis: computer programming and analysis
Scholarship: 3.2 grade-point average over the four years

Related Experience

9 months (part-time during senior year):
Assistant Programmer in the Business Service Office, California State University

3 months (full-time during summer 1980):
Assistant Programmer, ABC Records Corporation, Sacramento

References

Dr. James Roberts
California State University
6000 J Street
Sacramento, CA 95819

Mr. Paul Jones
General Manager
ABC Records Corporation
3100 E Street
Sacramento, CA 95823

Mr. Harvey White
Business Service Officer
California State University
6000 J Street
Sacramento, CA 95819

Letter in Response to a Newspaper Advertisement

3410 Carroll Drive
Center City, California 95002

July 1, 1981

Mr. Donald Gordon, Manager
KXXX Television
3814 High Street
Garden City, California 90003

Dear Mr. Gordon:

In reply to your advertisement in the Garden City Record, I would like to be considered for the news reporter's job at KXXX. I believe that my three years' experience at KXVO in Center City and my degree in broadcast journalism qualify me for the position.

I have been with KXVO as an on-camera news reporter since my graduation from California State University, Sacramento, in June 1978. I spend part of each day on the news beat, part of it writing stories and editing on-camera interviews, and part as co-anchor of our hour-long daily news broadcast.

One of my other duties at KXVO is to moderate a weekly panel show featuring prominent members of the community. I have provided more details of this task in my résumé.

At California State University, I had three years of broadcast experience on the campus radio station, KCCC, along with some administrative experience as broadcast news manager during my senior year. This taught me quite a bit about scheduling—both news broadcasts and reporters.

I believe my experience qualifies me for a position in a larger market, and I will appreciate an interview and audition at your convenience. I have enclosed a résumé listing my experience and education. I have discussed the opening at KXXX with our general manager, who agrees that I am ready for a news position in a larger station such as yours.

Very truly yours,

Robert F. Palmer

Robert F. Palmer

Enclosure: 1

Résumé in Seminarrative Style

Robert F. Palmer
3410 Carroll Drive
Center City, California
(912) 863-5424

Personal Information

Age: 27
Health: good
Marital status: married
Hobbies: contract bridge, golf

Education

California State University, Sacramento
Degree: B.A., Broadcast Journalism, June 1978
Scholastic record: During the four years, I maintained a 3.0 grade-point average.
Extracurricular activities: I joined the Journalism Club during my sophomore year and served as club president during my senior year.

Experience

KXVO, Center City, California, July 1978 to present:

1. I have three years' experience in three phases of television news: (a) the news beat; (b) writing stories and editing on-camera interviews; (c) co-anchoring a daily hour-long news program.
2. I also moderate a weekly panel show, on which community members discuss such items of local interest as law enforcement, library service, and public transportation. My responsibility is to coordinate the participants, ask pertinent questions, and keep the program moving.

KCCC, campus radio station at California State University

1. During my sophomore and junior years, I was a broadcast reporter and moderator of panel discussions.
2. During my senior year, I was appointed manager of the news department. My principal responsibility was to schedule the news broadcasts and the reporters.

APPOSITIVE. An appositive is a noun or noun equivalent placed next to another noun to amplify or explain it: for example, Kepler *the astronomer*. As you can see from the example, the appositive (*the astronomer*) has the same meaning as the word it modifies and in the preceding example is restrictive. See **Restrictive and Nonrestrictive Modifiers.**

Here are other examples of *restrictive appositives*:

1. President *Jones* (appositive) will speak tonight.
2. His sister *Mary* (appositive) will graduate in June (restrictive because he has another sister).
3. The planet *Jupiter* (appositive) has eleven satellites.

Nonrestrictive appositives are set off with commas:

1. Our president, *Thomas Jones,* will speak tonight.
2. Jupiter has eleven satellites, *or moons.*
3. The original theatre, *a 100-year-old wooden structure,* will be restored.

The nonrestrictive appositive is a handy device for defining a term that may need a brief explanation or definition:

1. The communitor, *or grinder,* chops up large pieces of organic material, such as tree limbs and even small trees.
2. Primary, *or the first stage of,* waste-water treatment removes floating materials and settleable solids.
3. The air-stripping process removes nitrogen, *a nutrient that stimulates the growth of algae and other aquatic plants.*

When used in nonrestrictive constructions, appositives must be set off with commas so that readers are not misled into thinking they are reading about two things. In the first sentence, for example, the writer does not mean "either the communitor or the grinder." He simply used the appositive to remind the reader that the communitor is a grinder.

Properly used, the nonrestrictive appositive can be helpful to both readers and writers. In the second sentence above, the addition of the appositive enables the writer to remind the reader in just five words that "primary" is the first stage of waste-water treatment. In the third example, the writer used a subordinate clause to explain that nitrogen stimulates the growth of algae in water.

APPRAISE, APPRISE. *Apprise* means to inform. *Appraise* means to evaluate or judge. For example:

1. Please keep me *apprised* of your progress.
2. The building will be *appraised* next week.

AREA, IN THE AREA OF. Avoid the jargonish use of *area* to mean *category*. The phrase most often is meaningless. For example:

Jargon: *In the area of* science, I have studied physics and chemistry.
Direct: My natural sciences included physics and chemistry.

Jargon: *In the area of* construction, they have ten years' experience.
Direct: They have ten years of construction experience.

Jargon: *In the area of* automobile sales, the "X" cars are selling well.
Direct: The "X" cars are selling well in today's automobile market.

AROUND, ABOUT. See **About, Around.**

AS PER. *As per* is a touch of military or legal jargon and is too stuffy for modern writing. For instance, *as you requested* is more graceful than *as per your request*. Similarly, *as directed* is more natural than the somewhat pompous *as per the instructions*. In short, *as per* has seen its day.

AS REGARDS, WITH REGARD(S) TO. Both are inflated variants for *about:*

Inflated: The city council asked several questions *with regard to* the proposed parking garage.
Inflated: *As regards* the new garage, the city council had several questions.
Improved: The city council asked several questions *about* the new parking garage.

ASTERISK (*). Asterisks can be used to indicate footnotes, although consecutive numbers are often more convenient. When more than one footnote appears on a page, a writer has the problem of double asterisks, triple asterisks, etc. (Printers can use asterisks, daggers, and other marks, but the asterisk is the only such mark on a standard typewriter.) When an asterisk is used, it is raised one-half space following the statement needing a reference and repeated at the beginning of the footnote. (See **Documentation,** *Footnotes*.) Note, incidentally, the spelling and pronunciation: it is not an *asterick*.

AS TO, AS TO WHETHER. *As to* is an awkward substitute for *on* or *about; as to whether* merely means *whether*. For instance:

Awkward: His statement *as to* the parking problem was clear.
Improved: His statement *on* (or *about*) the parking problem was clear.

Awkward: I'm not certain *as to whether* we can finish by March 1.
Improved: I'm not certain *whether* we can finish by March 1.

B

BALANCE, REMAINDER. The use of *balance* to mean *remainder* or *rest* is colloquial and considered substandard. For instance:

Loose: The *balance* of the crew will arrive tonight.
Precise: The *rest* of the crew will arrive tonight.
Precise: Most of this season's snow fell during December and January; the *remainder* (not *balance*) fell in February and March.

BALANCED SENTENCE. In a balanced, or parallel, sentence, similar ideas appear in similar form—adjectives, nouns, phrases, clauses, etc. When similar ideas are presented in dissimilar form, readers may have to stop and search for the meaning. For example:

The book contains sections on grammar, punctuation, and also discusses words.

In balanced form, the sentence would clarify the three like ideas:

The book contains a section on grammar, a section on punctuation, and a section on words.

In the revised version, the three similar ideas arc presented in three similar noun phrases.

For additional discussion and examples, see **Sentence Types and Construction.**

BASED ON. This expression is often incorrectly used in place of *on the basis of. Based* is the past tense (and past participle) of the verb *base* and should be used as a verb. For instance:

Right: The March report was *based on* February snowfall data.
Wrong: The report was prepared *based on* February snowfall data.
Right: The report was prepared *on the basis of* February snowfall data.

BECAUSE. Avoid redundancy with *because,* a **conjunction** that introduces a subordinate **clause** explaining the main clause of a sentence.

Redundant: The reason why he failed the test was because he did not study.
Improved: He failed the test because he did not study.

Synonyms for *because* include *since, as,* and *for*. However, none of them is as exacting and all can sometimes be misleading. For instance: "We hurried home as the snow was falling." (Did we hurry home *in the snowfall* or *because* the snow was falling?)

Several inflated variants, inherited from conversation, are also used in place of *because*. These include *for the reason that, due to the fact that, by virtue of the fact that, the reason being that*. In technical writing, such expressions are best forgotten, particularly when they can be reduced to one precise word. See also **Inflated Expressions.**

BI (as a prefix). See **Prefixes and Suffixes.**

BIANNUAL, BIENNIAL, BIMONTHLY, BIWEEKLY. The correct use of these terms requires careful distinction between them. Although the prefix *bi* generally means *two,* these terms have acquired special meanings and should be used correctly:

1. Both *biannual* and *semiannual* mean twice a year.
2. *Biennial* means once every two years.
3. *Bimonthly* and *biweekly* mean once every two months and once every two weeks, respectively.
4. *Semimonthly* and *semiweekly* mean twice a month and twice a week, respectively.

BIBLIOGRAPHY. See **Documentation.**

BORN, BORNE. Both of these homonyms are past participles of the verb *bear*. The proper use for *born* is in the sense of being born ("He was born in Oregon"). *Born* is also used figuratively as an adjective: a born liar, a born actor. In the sense of *gave birth to* or *carried,* the correct form is *borne*:

1. His wife had *borne* four children.
2. *Borne* by a strong north wind, the kite was almost out of sight.

BOTH . . . AND. These **correlative conjunctions** are frequently helpful for eliminating ambiguity:

Confusing: An evaluation of the latest test results and the progress to date indicates that the program will succeed. (Were both the test results and the progress evaluated?)

Clear: Evaluations of both the latest test results and the progress to date indicate . . .

See also **Correlative Conjunctions.**

BRACKETS. Brackets are primarily editorial marks and are not found on the standard typewriter keyboard. They have three principal uses:

1. To set off parenthetical words within parentheses:

   The explanation can be found in all standard textbooks. (See, for example, Johnson's *Essentials of Arithmetic* [2nd edition], page 16.)

2. To enclose words inserted into a quotation:

   The economist stated, "Vegetables will continue in short supply [probably because of the drought] throughout the year."

3. To enclose the Latin word *sic,* which is inserted in quoted material to indicate an error in the original quotation:

   The story stated, "The kidnappers are believed to have crossed the Mexican boarder [sic]."

BRAINSTORMING. Brainstorming can help you obtain ideas for a report or other paper. Traditionally, it has been used at group conferences to obtain solutions to problems. Under the rules for brainstorming, every suggestion is acceptable as it is made; the evaluation and culling of unworkable ideas come after all the ideas are in.

To brainstorm a topic, jot down every idea that comes into your mind, with no thought of its merit or organization. Then, when you have exhausted the subject, you will probably have far more ideas and topics than you can use. Of course, brainstorming a topic may take you several hours, and you will still have to evaluate every idea and then organize those you decide to keep.

It is far better, however, to spend some time picking your brain than to spend hours staring at the walls, wondering what to write and how to start. Brainstorming will prime your engine and at least get you off the ground.

See also *Methods of Organization*, part 1, pp. 8–13.

BRAKE, BREAK. After receiving a student paper describing his job as a "break" [sic] mechanic—in which he also discussed the "breaking" system and used the wrong word eight times—I suggested that perhaps he had better put on the *brakes* and *break* open a dictionary. (On the other hand, I was forced to admit that he was at least consistent.)

BREADTH, BREATH. *Breadth* means width or extent:

1. I was amazed by his *breadth* of knowledge.
2. He took a deep *breath.*

BULK, THE BULK OF. Avoid using *the bulk of* when you mean *most*:

Loose: We will ship *the bulk of* the parts next month.
Precise: We will ship *most* of the parts next month.

Loose: The bulk of the snow fell during December and January.
Precise: Most of the snow fell during December and January.

BUSINESS LETTERS. See **Correspondence.**

BY MEANS OF, BY USE OF. These **prepositional phrases** are often redundant; *by* or *with* will usually suffice:

Inflated: Remove the bolts *by use of* a socket wrench.
Improved: Remove the bolts *with* a socket wrench. (OR)
Improved: Use a socket wrench to remove the bolts.

See also **Inflated Expressions.**

C

CANNOT, CAN NOT. Either expression is acceptable; the second form is sometimes used to emphasize the negative thought:

1. I *cannot* finish my report by Friday.
2. You *can not* violate the contract.

CAPITAL, CAPITOL. You can easily distinguish between these two if you merely remember that *capitol* has but one meaning: the *building* that houses the United States Congress or a state legislature. In every other usage, the word is *capital.*

1. The *capitol* (building) is in Boise. The *capitol* grounds are well landscaped.
2. Austin is the *capital* of Texas.
3. He lived in the *capital* city for many years.

4. We cannot raise the required *capital*.
5. He is opposed to *capital* punishment.

CAPITALIZATION. The use of capital letters at the beginning of sentences and for proper names is a convention followed by all writers. Certain other usages reflect divided opinions or are matters of taste. Many governmental organizations and business concerns establish particular conventions for the use of capital letters. In such cases, writers are expected to follow the conventions.

The following conventions apply to most writing situations.:

1. Capitalize the names of streets, parks, buildings, etc.

Madison Avenue	Capitol Park	Waco City Hall
Fifth Street	Resources Building	Fair Oaks Library
Field Museum	Merchandise Mart	Capitol Mall

   However, none of those words would be capitalized when used without a proper name. For example, "We are going to the library and then to the city hall."
2. Capitalize these words when they are written in the singular and follow a single proper name:

Canal	Delta	Ocean	River
Canyon	Forest	Pond	Sea
Creek	Lake	Range	Strait

   For example:

   a. The American *River;* BUT the American and Sacramento *rivers.*
   b. His home is in Blue *Canyon;* BUT The river runs along the bottom of a *canyon.*
   c. The Sacramento–San Joaquin *Delta*; BUT Near the sea, the river forms a wide *delta.*
   d. The Black *Forest;* BUT a dense *forest.*

3. Capitalize words denoting political subdivisions—county, city, state, precinct, province, territory, kingdom, etc.—when the word follows the name and is considered part of it or a substitute for it. However, when such words precede the name or stand alone, they are usually (but not invariably) written in lowercase. Some examples:

   a. New York *State;* BUT the *state* of New York.
   b. Placer *County;* BUT the *county* of Placer.

c. New York *City;* BUT the *city* of New York.

d. Fourth *Precinct;* BUT the *precinct.*

e. The *Republic* (when referring to a particular country, such as the United States, France). For example, "The *Republic* is sound." BUT France is a *republic.*

f. Northwest *Territory;* BUT friendly *territory.*

g. *Province* of Quebec; BUT Quebec is a *province.*

4. Capitalize a title preceding a personal name; however, when titles follow a name or stand alone, they usually are not capitalized:

   a. *President* George Washington; BUT George Washington, first *president* of the United States.

   b. He is the *president*; the *president* will speak tonight.

   c. *Governor* Edmund G. Brown, Jr.; BUT the *governor* of California.

   d. *Chief Justice* Warren Burger; BUT Warren Burger is the *chief justice.*

   e. *Pope* John; BUT the *pope.*

   f. *King* George VI; BUT the *king* of England.

5. Capitalize the days of the week and months of the year. However, the four seasons are written in lowercase (unless personified): Monday, July, spring, winter; BUT Old Man Winter arrived with an icy blast.
6. Capitalize compass directions when they designate a particular region or a special name for a region:

   the South — Southern California (regional name)
   the West — Upper Michigan (regional name)
   the Near East — the Midwest

   However, do not capitalize "directional" parts of states or countries:

   eastern Utah — northern Idaho
   the midwestern states — western Colorado
7. Capitalize divisions, departments, and offices of government when the official name is used:

   Department of Finance — California Legislature
   Congress (U.S.) — California Assembly
   Senate (U.S.) — Superior Court of Ada County
   House of Representatives
   Environmental Protection Agency
8. Capitalize the titles of academic degrees when using them in full:

Master of Arts — Bachelor of Science
Doctor of Philosophy — Doctor of Divinity

However, when referring to degrees in general terms, do not capitalize them: doctorate, master's degree, bachelor's degree. Some more specific examples:

a. John Roberts, *Bachelor of Science* (or B.S.).
b. Jane Andrews, *Doctor of Philosophy* (or Ph.D).
c. He has a *master's* degree.
d. She is working toward a *doctorate.*
e. He hopes to complete a *Master of Science* degree in June.

CASE. The case of a **noun** or **pronoun*** depends on its function in a sentence. A noun may be the **subject** (nominative, or subjective, case); it may be the **direct object** of a verb or preposition** (objective, or accusative, case); or it may be in the possessive, or genitive, case. Here are some examples:

Nominative: The *boy* (subject) hit the ball.

Objective:

1. *Direct object of a verb:* The boy hit the *ball.*
2. *Object of a preposition:* They walked along the *river.*

Possessive:

1. *Shown by apostrophe and "s":* The *boy's* award was well deserved.
2. *Shown by "of":* The format *of* a report is important.

Note: Use *of* to show the possessive case of inanimate objects rather than an apostrophe and *s.* In some cases, a noun **modifier** is preferable to the possessive. For instance, "the *automobile* engine" (rather than "the automobile's engine" or "the engine of the automobile").

See also **Apostrophe.**

CATALOG, CATALOGUE. Take your choice with this pair; the shorter form is simpler and is preferred by most modern writers and organizations.

CAUSE AND EFFECT. See **Analysis, Organization of.**

*The case of pronouns is discussed on p. 203.
See also **Indirect Object.

CIRCUMLOCUTION. Circumlocution is roundabout or indirect expression—the use of excessive words and phrases. See **Inflated Expressions.**

CITE, SIGHT, SITE. To distinguish between these three, keep in mind the following sentence: "While searching for a building *site* away from the *sights* of the city, Mr. Brown was *cited* for trespassing."

CLASSIFICATION, ORGANIZATION OF. To classify is to arrange facts in categories, a most useful pattern of organization for technical and scientific writers. When a writer classifies facts or items, readers can usually determine the relationships between those facts most readily. For example, a writer might classify the bodies that occupy the solar system as planets, planetoids (asteroids), and comets. A botanist might classify all plant life into four main divisions: (1) thallophytes, (2) bryophytes, (3) pteridophytes, and (4) spermatophytes. Botany itself can be classified into four main disciplines, or divisions. Here is a brief explanation and classification of the science of botany that clarifies the four major botanical disciplines: morphology, physiology, ecology, and systematics.

Morphology. Morphology deals with the structure and form of plants and includes such subdivisions as cytology, the study of the cell; histology, the study of tissue; anatomy, the study of the organization of tissues into the organs of a plant; reproductive morphology, the study of life cycles; and experimental morphology, or morphogenesis, the study of development.

Physiology. Physiology deals with the functions of plants. Its development as a subdiscipline has been closely interwoven with the development of other aspects of botany, especially morphology. In fact, structure and function are sometimes so closely related that it is impossible to consider one independently of the other. The study of function is indispensable for the interpretation of the incredibly diverse nature of plant structures. In other words, around the functions of the plant, structure and form have evolved. Physiology also blends imperceptibly into the fields of biochemistry and biophysics, as the research methods of those fields are used to solve problems in plant physiology.

Ecology. Ecology deals with mutual relationships and interactions between organisms and their physical environments. The physical factors of the atmosphere, the climate, and the soil affect the physiological functions of the plant in all its manifestations, so that, to a large degree, plant ecology is a phase of plant physiology under natural and controlled conditions; in fact, it has been called "outdoor physiology."

Plants are extremely sensitive to the forces of the environment, and both their association into communities and their geographical distribution are determined largely by the character of climate and soil. Moreover, the pressures of the environment and of organisms upon each other are potent forces, which lead to new species and the continuing evolution of larger groups.

Systematics. Systematics deals with the identification and ranking of all plants; it includes classification and nomenclature (naming) and enables the botanist to comprehend the broad range of plant diversity and evolution.[3]

CLAUSES AND PHRASES. A *clause* is a sentence element containing a **subject** and a **verb.** Clauses are classified as independent and subordinate (dependent); an independent clause makes sense by itself, but a subordinate clause depends on an independent clause, both grammatically and for meaning. For example:

Independent: He was late for work yesterday.
Subordinate: Because his car would not start, he was late for work.

Note how the subordinate clause "depends" on the independent clause: *Because his car would not start* is grammatically incomplete and, without the main clause, has no complete meaning.

A *phrase* is simply a group of words used as an adjective, adverb, noun, or verb. The principal grammatical difference between a phrase and a clause: a phrase contains no **subject and predicate** but serves simply as a **modifier,** a **noun,** or a **verb.**

Prepositional phrase (adjectival): He is a man *of high principles* (modifies the noun *man*).

Prepositional phrase (adverbial): She works *in the main laboratory* (modifies the verb *works*).

Noun phrase (gerund): Playing golf is good exercise.

Verb phrase: The metric system *has been used* for many years.

Participial phrase: Entering the cave, I heard water running.

Absolute phrase: Our assignments finished, we went to the ball game.

See also **Sentence Types and Construction.**

CLICHÉ. Clichés are trite expressions that have been used so often they have lost their "zing." Such overworked expressions as "sound as a dollar," "in the spotlight," "center stage," and "passed the test with flying colors" are best forgotten unless you can come up with a novel way of using them. Actually, such expressions are seldom applicable to technical writing,

[3]"Botany," *Encyclopaedia Britannica*, 15th edition, 1974. Reprinted by permission.

although I have seen all four of the preceding quoted phrases in recent articles and reports.

A number of trite expressions that have been used for many years in business correspondence are now considered clichés. A few of these include "Hoping to hear from you soon, I remain"; "The undersigned regrets to state"; "It has come to the attention of the writer"; "Please don't hesitate to contact us." (For a discussion of clichés in **correspondence,** see page 85.)

Actually, nothing is intrinsically wrong with trite expressions and clichés; they have simply lost all power to capture a reader's imagination. Once in a while an old saying may be useful. Consider, for example, this advertising slogan of a California morning newspaper, the *Sacramento Bee*: "Northern California wakes up to the birds and the *Bee*." Unless you can do as well, however, stay away from clichés and tired expressions.

CLOSE PROXIMITY. This is a simple redundancy that writers should avoid. *Proximity* denotes closeness or nearness; therefore, the overused prepositional phrase *in close proximity* can be reduced to *close* or *near.* Government writers are especially given to describing things as *located in close proximity* to other things. For instance, "The sewage plant is *located in close proximity* to the river." This simply means, "The sewage plant is *near* the river." (Note that the word *located* is also superfluous; *located* is implicit in the phrase *is near the river.*)

CO (as a prefix). See **Prefixes and Suffixes.**

COHERENCE. Writing is said to be coherent when its meaning is clear to a majority of those who read it. Coherence is an elusive quality, however, and may be difficult to attain for a writer who fails to consider readers' needs.

Accordingly, one of the best ways to achieve coherence is to put yourself in the readers' place and examine your writing from their viewpoint rather than your own. Consider that you, the writer, are at the end of the particular experience under discussion but that the readers are just starting. You have all the facts and figures, and now you must present them in a way that "outsiders" can understand.

Another way to achieve coherence is to recognize—and avoid—the snares that lead to incoherence, many of which are discussed in this book. Here are eleven of the most prominent:

1. Poor organization of information. See *Organization*, part 1, pp. 6–13.
2. Poorly constructed sentences. See **Sentence Types and Construction.**
3. Poorly connected thought relationships. See *Provide Transition Within Paragraphs* and *Provide Transition Between Paragraphs*, part 1, pp. 29–32.
4. Careless use of modifiers. See *Be Careful with Modifiers*, part 1, pp. 20–21.
5. Careless reference of pronouns. See **Sentence Types and Construction.**
6. Careless punctuation. See **Punctuation.**
7. Use of vague words and expressions. See *Abstract and Concrete Words*, part 1, p. 15.
8. Use of jargon or technical words. See *Avoid Highly Technical Words*, part 1, p. 15, and **Technical Terms and Jargon** (Handbook, p. 256).
9. Use of incorrect words. See *Use The Correct Word*, part 1, p. 16, and **Homonyms; Malapropisms** (Handbook, pp. 150 and 166).
10. Use of too many words. See *Watch for Redundancy and Circumlocution*, part 1, p. 17, and **Inflated Expressions** (Handbook, p. 154).
11. Use of too few words. See *Avoid Superconcise Writing*, part 1, p. 19, and **Prepositional Phrases** (Handbook, p. 196).

A singular problem many writers face is their intimate knowledge of the subject at hand. This would appear to be advantageous; however, even knowledgeable writers may forget that their readers often know less than they do about a particular subject. Accordingly, many writers may move too quickly through a complicated discussion, they may use technical terms without defining them, or they may omit basic explanations that readers often need.

Contrast the writer's task, for a moment, with that of a speaker. Speakers, of course, must also present information coherently. But they also have certain advantages that writers do not enjoy. Speakers can use gestures, facial expressions, and word emphasis to help clarify meanings. Even more important, if listeners are confused, they can usually ask for immediate clarification.

Readers, unfortunately, don't often have opportunities to ask questions. They are at the mercy of writers and their communication skills. Therefore, to write coherently, writers must always consider their readers and how much they need to know.

COHORT, COLLEAGUE. *Cohort,* a collective noun denoting a group united in a common cause, should not be used to mean an individual or companion. In the latter sense, the correct word is *colleague, companion,* or *co-worker.*

COLLECTIVE NOUNS. See **Noun.**

COLLOQUIAL LANGUAGE. The term *colloquial* denotes language that is more suited to conversation than to writing, especially formal writing, and, in particular, business, technical, governmental, and "student" writing. Almost all of us talk more often than we write, and naturally our writing will be influenced by our speech.

One language that would appear to be uniquely applicable to both speech and writing is the peculiar lexicon of sports. Both the broadcaster and the sports writer can use such terms as *hurler* (pitcher), *feared slugger* (skillful batter), *bomb* (long pass), *hoop* (basket), *spikes* or *cleats* (shoes), *savvy* (knowledge of the game), and sports fans will understand and approve.

The conversational language of the business or governmental office, however, is usually considered inappropriate for business or governmental writing. In the office, for example, the *boss* is the *boss.* In most business writing, however, he or she would probably be referred to by some more formal title, such as *supervisor, manager, section chief.* Similarly, if a problem has you *stumped,* you would probably report in writing that you were *puzzled* or *confused.*

This distinction between formality and informality is also generally applicable to oral and written reporting. Today, most speakers lean toward conversational language and in most cases speak more casually than they write. Speakers, for example, prefer the conversational *it's, they're, he'll,* etc., whereas most writers of reports tend to shun contractions and use the more formal *it is, they are, he will.* A few other variations between the spoken and written language follow:

Spoken	Written
a lot, a lot of	much, many, considerable, a large number of
busted	broken or damaged
chance it	take a chance
go along with	agree with, concur in
go overboard	go too far, overextend
get acquainted with	learn

hard	difficult
hassle	argue, quarrel
fix	repair
sloppy	untidy
it's, he'll, we're (most contractions)	it is, he will, we are
OK	all right, acceptable, satisfactory, suitable

Unfortunately, there are no rules governing either looseness or formality in written communication. Here is another situation that depends on a writer's judgment. A sound rule is to write with some degree of formality without sounding stuffy or pompous.

In many cases you may have to rely on trial and error. If, for example, a colloquial expression seems too informal to you, try to find a more appropriate synonym. If, however, the substitute seems too stuffy, perhaps the colloquial expression will be more suitable after all. Another good test is to read a questionable passage out loud. Then ask yourself, "How would I react if I had received this from somebody else?"

Finally, colloquial language, however informal, differs from **slang,** the vocabulary of which comprises mainly unconventional and transitory words and expressions—*cool, far out, right on, endsville.* Suffice it to say that slang has no place in technical, business, or governmental writing.

Compare with **Pompous Language; Slang.** See also **Get, Got; Good; Inflated Expressions.**

COLON. The colon is a mark of anticipation or introduction, directing a reader's attention to what lies ahead. Use it:

1. *After an introductory statement:* "The system operates as follows: the pressure circuit energizes . . ."
2. *To introduce a series or enumeration:*

   a. The schematic is divided into two circuits: a control circuit and a power circuit.

   b. Lumber was supplied by the following contractors: (1) Johnson and King, (2) Roberts and Smith, and (3) Thompson and Ziegler.

3. *Between two clauses of a sentence* when the second is an amplification of the first or a restatement in different terms: "The X-3 motor has not been perfected: it must still be flight-tested."

Note: Don't interrupt a simple sentence with a colon:

Wrong: We need: pens, pencils, ink, and envelopes.
Right: We need the following supplies: pens, pencils, ink, and envelopes.

COMMA. The comma is probably the most abused punctuation mark of all. Its principal function is to indicate a pause or a necessary separation of words. Whenever you are uncertain of the need for a comma, try the sentence out loud; often you will be able to "hear" whether it belongs. Use a comma:

1. Between the two parts (**clauses**) of a compound sentence (a sentence containing two or more independent clauses): "He attended a community college for two years, and then he went on to the state university."
2. To separate the two parts of a complex sentence (a sentence containing at least one independent clause and one subordinate clause): "Although I was late for class, I managed to finish the examination."

   Caution: Inexperienced writers often insert a comma following the subject when no comma is needed. For instance: "The rocket engine, was installed in the test stand." This is definitely a taboo. Read that sentence out loud, and your ear will tell you that the comma following *engine* is superfluous.
3. To prevent a misinterpretation:

   a. After the visitors left, the tank was completely drained. (NOT) After the visitors left the tank, etc.

   b. Inside, the rooms were cold and dark. (NOT) Inside the rooms, etc.

4. To set off an introductory phrase or **conjunctive adverb:**

   a. For example, units of measure are often abbreviated.

   b. Consequently, the shipment was delayed.

5. After each **adjective** in a series of dependent **modifiers,** except for the one immediately preceding the modified word. (If modifiers are dependent, they will make sense with *and* between them or in reverse order.) For instance: *Dependent:* a long, narrow room (a long *and* narrow room). *Independent:* a heavy metal container (not a heavy *and* metal container).
6. Between all units of a series, including the word before the *and* that precedes the final element: "Corn, wheat, and barley are bringing higher prices this year."

   Opinions differ on the use of a comma before the *and* in such constructions. But now consider this sentence: "Three pleasing color combinations are red and blue, orange and yellow, and brown and gold." Note what would happen to a reader if the final comma were omitted. Accordingly, if you always use one before the final *and,* you will have less worry about confusing a reader.

When any of the series elements contains an internal comma, it may be wise to separate the elements with **semicolons:** "The replacement kit contains nuts and bolts, which are used to replace the original rivets; spare wiring, which must be cut to fit; and complete instructions, which must be followed carefully." See also **Semicolon.**

7. To separate the parts of dates and places: January 15, 1980; Columbia, South Carolina.

COMMA SPLICE. Never join two independent clauses with a comma, or you will create a comma splice and run-on sentence. For example:

Wrong: They are coming tomorrow, we can get the answer then.
Right: 1. They are coming tomorrow; we can get . . .
2. They are coming tomorrow, and we can get . . .
3. They are coming tomorrow. We can get . . .

See also **Sentence Types and Construction.**

COMPARATIVE AND SUPERLATIVE OF ADJECTIVES AND ADVERBS. Writers should be aware of three points concerning comparisons:

1. To show a greater (or lesser) degree of what is being named, change **adjectives** and **adverbs** as follows:
   a. The addition of the suffixes *er* or *est*
      (1) *Adjective*

      Comparative: John is *taller* than Henry. (Of the two, John is the *taller,* not the *tallest.*)
      Superlative: John is the *tallest* boy in the class.

      (2) *Adverb*

      Comparative: John can run *faster* than Bill can. (Of the two, John is the *faster* runner, not the *fastest.*)
      Superlative: John runs the *fastest* of anyone on the team.

   b. The addition of *more* or *most*
      (1) *Adjective*

Comparative: Henry's car is *more* luxurious than John's. (Of the two, Henry's car is the *more* luxurious, not the *most* luxurious.)
Superlative: Henry's car is the *most* luxurious of the entire line.

(2) *Adverb*

Comparative: John handles the boat *more* skillfully than Henry does.
Superlative: John handles the car the *most* skillfully of all the drivers.

Note that when only two persons or things are compared, the correct suffix is *er,* not *est;* the correct adverb with two persons or things is *more,* not *most.* Of two test scores, for example, one will be the *higher,* not the *highest;* of three, of course, one will be the *highest.*

2. Make logical comparisons and use consistent phraseology.

Illogical: The data by Johnson show more consistent results than Smith. (Illogical because the statement compares things that cannot be compared—*data* and *Smith.*)
Logical: Johnson's data show more consistent results than Smith's. (OR) The data by Johnson show more consistent results than those by Smith.

Illogical: Her writing is better than the usual college student. (Again an illogical statement, comparing *writing* and *student.*)
Logical: Her writing is better than the usual college student's.

Inconsistent: Her writing is as good if not better than the usual college student's. (Inconsistent because the phrase *as good* is incomplete; the correct idiom is *as good as.*)
Consistent: Her writing is as good as if not better than the usual college student's.

3. Be careful with comparisons of such "absolutes" as *certain, complete, final, perfect, unique.* Strictly speaking, there are no degrees of certainty, completion, finality, perfection, or uniqueness. Linguists, however, disagree on the appropriateness of such statements as "The summary could be *more complete,*" "I am *more certain* than I was last month," and "The countryside is *most unique.*"

The usage panel of the *American Heritage Dictionary* defends the use of *more certain;* accepts *more complete* when the meaning is *more thorough;* remains silent on *final;* accepts *more perfect* in the sense of excellence; and frowns on the use of comparatives with *unique,* which means "the only one of its kind." Other linguists offer varying opinions; most often cited is the Constitution's "more perfect union."

The safer course would be to avoid illogical comparisons of such terms, unless you use them in a figurative sense or for emphasis. For instance, "If you don't complete your laboratory work, you'll *most certainly* have to repeat the course." For further discussion, see Theodore Bernstein's *The Careful Writer;*[4] Roy Copperud's *American Usage and Style;*[5] and an unabridged or "College" edition of the *American Heritage Dictionary of the English Language.*

See also **Unique.**

COMPARISON AND CONTRAST. Comparison and contrast are often used to show similarities and differences between two or more things, persons, places, institutions, processes, etc. You might compare two processes, such as evaporative and nonevaporative cooling, to point out that one is more efficient or economical in certain climates.

You might also use comparison and contrast to introduce an unfamiliar process by showing how it resembles or differs from a familiar process. For instance, you might explain the principle underlying evaporative cooling by reminding readers of the chill they feel when emerging from a warm bath or shower, an experience everyone has had. Then you could further explain how water evaporating from the human body takes heat from it, and thus cools it, to introduce an explanation of an evaporative-cooling mechanism, such as a cooling tower at a nuclear power plant.

Comparison and contrast are often used to clarify expanded **definitions.** The definition on page 105 compares two statistical terms—population and sample.

The author of the next example used comparison and contrast to introduce the three main printing processes. In the discussion, the author italicized certain words that emphasize the differences between the three processes:

> The three major printing processes are letterpress (printing from a raised surface), offset-lithography (printing from a level, or plane, surface), and gravure (printing from a depressed, or sunken, surface).
>
> *Letterpress.* This is the oldest and, until recently, the most commonly used printing method. Ink is applied to a *raised* surface and transferred directly to the paper through pressure. The areas to be printed are *raised* above the nonprinting areas; the ink rollers touch only the top surface of the raised areas. The surrounding (nonprinting) areas are lower and do not receive ink.

[4]New York: Atheneum, 1965, p. 228.

[5]New York: Van Nostrand, 1979, p. 82.

C

Offset-lithography. This is the newest of the three processes as well as the fastest growing. Lithography consists of printing from a *plane,* or flat, surface that is neither raised nor depressed. The printing image is level with the nonprinting areas surrounding it. Two basic differences between offset and other processes are: (1) the use of the lithographic principle that grease and water do not mix, and (2) the manner in which ink is placed on the paper, *offsetting* it first from plate to rubber blanket and then from blanket to paper.

A major advantage of this process is the clearer impression created by the soft rubber surface (as compared to letterpress metallic plates) on a wide variety of papers and other materials. As commonly used today, *offset* refers to the *process* that transfers an image to paper by means of three cylinders, instead of two as with letterpress. The inked and watered plate prints on a rubber blanket, which in turn *offsets* this ink impression to the paper. Printing from a rubber blanket eliminates much of the makeready time (preparation of heavy and light areas so that both areas print with the correct impression) that characterizes letterpress printing.

The offset process differs from letterpress in other ways. For example, instead of assembling type and plates, *photographic negatives* or *positives* are used to make the plates. One printing plate is required for each color. Letterpress forms and plates can be readily converted for use in offset-lithography by one of several conversion systems.

Gravure. Whereas letterpress uses a *raised* surface and offset a *flat* surface, gravure uses a sunken, or *depressed,* surface for transferring the image. A copper wrap-around plate, or cylinder, with the image etched *below* its surface rotates in a bath of ink. The excess is wiped off by a *doctor blade.* The ink remaining in thousands of recessed cells, or *wells,* forms the image by direct transfer to the paper as it passes between the plate and impression cylinders.[6]

COMPLEMENT, COMPLIMENT. A *compliment* is a word of praise or admiration. A *complement* is one of two parts that mutually "complete" a whole. For instance, two *complementary* angles complete a 90-degree angle. Complementary colors are two colors that blend together to make up a third color (as blue and yellow make green).

Complement also means an amount that completes a whole. For example, "The laboratory has a full *complement* of technicians."

COMPLEX SENTENCE. See **Sentence Types and Construction.**

COMPOUND NOUN. See **Hyphen.**

[6]Reprinted with permission of International Paper Company from the 9th edition (1968) of its copyrighted publication, *Pocket Pal: A Graphic Arts Production Handbook.*

COMPOUND PREDICATE. Two or more verbs with the same subject make up a *compound predicate.* Don't use a **comma** after the first clause, or you will separate the second verb from the subject. For example, "The manager approved the plans and set the starting date for construction."

However, an intervening phrase is set off with commas: "The manager approved the plans and, after reviewing the data, set the starting date for construction."

See also **Subject and Predicate.**

COMPOUND SENTENCE. See **Sentence Types and Construction.**

COMPOUND SUBJECT. Two or more nouns written as the subject of one verb make up a *compound subject.* The verb is usually plural; however, when the second noun is connected to the first by *together with, along with,* or *as well as,* the construction is usually singular. Singular subjects connected with *or* also take a singular verb. For example:

1. The president and the committee are leaving for New York.
2. The president, along with the committee, is leaving for New York.
3. Either the president or the vice-president is going to conduct the meeting.
4. Neither the president nor the vice-president is here.

See also **Sentence Types and Construction** and **Subject and Predicate.**

COMPOUND WORDS. See **Hyphen.**

COMPRISE. Strictly speaking, *comprise* means *consists of* or *includes* and should not be used in the passive construction *is comprised of.* A simple rule: the whole *comprises* (consists of) the parts; the parts *make up* the whole. For example:

Wrong: The study program *is comprised of* eighteen lessons.
Wrong: Eighteen lessons *comprise* the study program.
Right: The study program *comprises* eighteen lessons.

CONCISENESS. You are writing concisely when you use a minimum number of words to express your meaning clearly and precisely. See *Write Concisely (But Not Too Concisely),* part 1, p. 17; **Absolute Phrase; Expletive; Inflated Expressions; Prepositional Phrase.**

CONJUGATION OF VERBS. See **Verb.**

CONJUNCTION. As the name implies, a conjunction is used to join **clauses and phrases** and may also be used to introduce a clause:

Connective: I was late for work *because* my car would not start.
Introductory: Because my car would not start, I was late for work.

Conjunctions are generally classified as *coordinating* (and, but, for, etc.) and *subordinating* (because, although, so, when, so that, such as, etc.):

1. A coordinating conjunction is used to connect sentence elements (words, phrases, clauses) of equal rank (coordinate elements). For example:

   Clauses: He worked very hard, *and* he really earned his promotion (two independent clauses).
   Phrases: across the river *and* through the trees (two prepositional phrases).

2. A subordinating conjunction is used to connect a subordinate (dependent) clause to an independent clause:

   He missed a week of school *because he was ill.*

See also **Correlative Conjunctions.**

CONJUNCTIVE ADVERBS. See **Adverb.**

CONNOTATION, DENOTATION. The *denotation* of a word is its literal meaning. When you look up a word in the dictionary, you will find an explicit definition. The same word, however, may have several *connotations,* or implications; many words can suggest meanings or ideas beyond their dictionary definitions.

For instance, suppose you write, "John's work is generally satisfactory." Here, the use of *satisfactory* denotes acceptability, that John is doing acceptable work. It could also imply (connote) that his work is just barely passable, is just above average, or could be much improved. The latter meanings are the word's connotations—the implications others may read into it.

Now suppose you write, "I never have to worry when John works on my car." This time you have again said that John's work is satisfactory, but you have also implied that his work is above average, is outstanding, or couldn't be better.

In another vein, the meanings of many ordinary words are often modified by how they are used. "He bought an *inexpensive* car" implies something different from "He bought a *cheap* car," or "He certainly bought that car *cheap*." Also, the word *inexpensive* will have different connotations for the person making $60,000 a year and the person making one-quarter of that salary.

Technical, business, and governmental writers can avoid creating the wrong connotations by using words that carry concrete, precise meanings. You certainly would not want to report that you had completed a series of "cheap" experiments. Rather, you would probably report that the experiments were inexpensive or that their average cost was just $20. Nor should you report that they were completed "in no time at all," when you can report that they were completed "in just thirty hours."

See also **Imply, Infer** (Handbook) and *Abstract and Concrete Words*, part 1, p. 15.

CONSENSUS OF OPINION. Avoid this common redundancy. *Consensus* means majority agreement or accord and thus "majority of opinion." For instance, "The consensus is against the death penalty."

CONSUL, COUNCIL, COUNSEL. The city *council* is the group that governs the town. The city *counsel* is the city attorney, who gives advice and counsel. A *consul* is a government official who lives in a foreign country and represents the citizens of his or her home country (as the United States consul in France).

CONTINUAL, CONTINUOUS. Strictly speaking, *continual* means occurring at intervals; *continuous* means constant or without interruption. For instance, the *continual* drip of a leaky faucet, but the *continuous* flow of water through a pipe. However, except in the sense of space (as a *continuous* stretch of highway), the two are often interchanged with no loss of meaning.

CONTRACTIONS. See **Apostrophe.**

CONVINCE, PERSUADE. Although both words mean to "influence another's thoughts or actions," to *convince* is to use logic or proof to influence opinion; to *persuade* is to influence another to take action. For instance:

1. He used statistics to *convince* us that he was right.

2. He presented a *convincing* argument for the reorganization plan.
3. He *persuaded* us to vote against the proposed parking plan.

COORDINATING CONJUNCTIONS. See **Conjunction.**

CORRELATIVE CONJUNCTIONS. Correlative conjunctions are used in pairs—the first introducing and the second connecting—to join sentence elements of equal rank. Because correlatives are coordinating conjunctions, the elements connected must be of equal grammatical value and in parallel form, e.g., two adjectives, two nouns, two phrases, two clauses. The principal correlatives are:

both . . . and	either . . . or	whether . . . or
neither . . . nor	not only . . . but also	

In each of the following examples, note that the words following both conjunctions are in identical form, thus making the statement balanced (parallel).

1. He reported that *both the ignition system and the carburetor* needed repair. (NOT the ignition system and the carburetor *both* needed repair.)
2. The motor must be *either repaired or replaced.* (NOT *either be repaired or replaced.*)
3. *Neither the ignition system nor the carburetor* is operating properly. (NOT *Neither* the ignition system *or* the carburetor . . .)
4. The new machine can be used *not only for cleaning but also for polishing.* (NOT . . . not only *for cleaning* but also *as a polisher.*)
5. *Whether the present laboratory is refurbished or new facilities are constructed,* the test program will continue. (NOT Whether the present laboratory is refurbished or we construct new facilities . . .)

CORRESPONDENCE. All writers should be aware of a significant difference between business and technical correspondence and other forms of exposition—reports, instructions, and other informative writing. The latter are usually quite objective, for example, impersonal accounts of past events, explanations to help readers understand something, or instructions to help readers accomplish a task. A letter, however, is usually from one person to just one other person and thus requires special writing techniques.

Of course, reports, instructions, business proposals, and the like are also intended for people but are usually impersonally written. In business proposals, the competence of certain staff members is often discussed, but, again, there is usually no personal contact between writer and reader. On the other hand, even though the subject of a letter may be impersonal—oil, steel, water, etc.—the letter represents a true person-to-person contact. In no other form of technical or business writing will you face the same personal communication problems. Therefore, you should be aware of certain special techniques for preparing business letters.

The substance of business correspondence is the subject of an entire course in many colleges, and the techniques required for writing different types of letters are treated only briefly here. The following paragraphs contain (1) nine guidelines for preparing effective letters, (2) brief discussions of the principal parts of letters, (3) forms for various inside addresses and salutations, (4) a list of two-letter state abbreviations, and (5) samples of several types of letters, including:

a. a letter asking a favor;
b. a letter requesting a personal recommendation;
c. a letter of complaint requesting an adjustment;
d. a letter refusing a request;
e. a letter granting an adjustment;
f. a letter refusing a request for adjustment.

Job application letters are discussed under **Application Letters and Résumés.**

Nine General Guidelines to Effective Letters

1. Business letters require special attention, because—unlike business reports, for example—they are brief communications from one person to another, and the writer must be careful to use words wisely.
2. Once a letter has been received, the writer and the organization he or she represents are on record. A letter cannot be rescinded, explained away, or apologized for. Therefore, a writer should (a) use standard English, (b) state facts accurately, (c) always be courteous, and (d) make certain that a letter is physically attractive.
3. Letter writers should watch carefully for misspelled words, careless use of words, and grammatical errors. Many persons will be offended by even small mistakes: the use of *principal* when the meaning is *principle,*

for instance, could seriously detract from the writer's image. In this respect, the brevity of most letters is important. Just one small error in a one- or two-page letter will be all too noticeable.

4. Accuracy and clarity are also important. If readers cannot understand a letter, they might decide to write for more information, or they might file the letter—in the wastebasket.
5. You should be neither too blunt nor overly loquacious. For instance, you can acknowledge receipt of a letter with a few gracious words. Instead of, "We have received your letter of July 3, and we are grateful for your prompt reply," try, "Thank you for your prompt reply of July 3," or some such succinct but courteous wording.
6. Avoid clichés and pompous expressions. A few hackneyed phrases that have shown up in letters for many years include the following:

   Pompous: We are in receipt of yours of May 5.
   Natural: Thank you for your letter of May 5.

   Pompous: Enclosed herewith is the information you requested.
   Natural: Here is the information you asked for.

   Pompous: As per your request.
   Natural: As you requested.

   Pompous: Please feel free to contact us.
   Natural: Please call or write.

   Pompous: Trusting you will find this settlement to your liking, I remain . . .
   Natural: I hope you will like our proposed settlement. We want you to be satisfied.

   The modern trend in business correspondence is to use plain language and a conversational tone. Stay away from clichés and tired phrases. (See also **Cliché; Pompous Language.**)
7. Use the "you attitude" whenever possible. This is simply an approach that suggests you are writing with the reader's interests in mind. For example, instead of, "*We* would like you to fill out the enclosed form so that *we* can process your application"—which sounds "we" centered—try, "*Your* application will be processed as soon as *you* have returned the enclosed form." Say YOU as often as possible, especially in the opening paragraph. To begin with "Thank you" for something the reader has sone preceding the letter will get you off to a fine start. There is more about the "you attitude" on p. 98.
8. A sound approach to effective correspondence: in writing any letter,

always be guided by the objective you hope to attain—closing a sale, obtaining a refund, asking a favor, etc.

9. When you have written a letter, read it carefully. Then ask yourself, "How would I feel if I had received this from somebody else?"

Principal Parts of a Business Letter

The principal parts of a business letter are the heading, the inside address, the salutation, the body, the complimentary close, and the signature. Some letters also contain an attention line.

Heading. The heading, which identifies the sender, contains three or four lines, including the date. The writer's name does not appear in the heading, although the name of an organization may be included. If the stationery bears a printed letterhead, the only line typed is the date. Here are two typical headings (typed):

4629 Spring Street
Harrison, Arkansas 72601
August 15, 1981

Western Boys Club
5326 Sumpter Street
Tuscon, Arizona 85710
September 23, 1981

Note that no abbreviations are used for any of the words in the heading.

The heading may be placed (1) to end at the right margin, (2) to begin just right of the center of the page, or (3) to begin at the left margin (sometimes called *flush left*). This last arrangement is in keeping with the modern trend of having every part of the letter begin at the left margin (*block* style).

Some organizations use a modified block style, in which all lines begin at the left margin except for the heading and the signature, which begin just to the right of the center of the letter. The sample letters at the end of this discussion demonstrate both styles. The block style is the predominant style today.

Inside Address. Letters addressed to persons should always bear a title, for example, Mr., Mrs., Dr., Professor. The only title other than Mr. or Mrs. that should be abbreviated is Dr. Never abbreviate *Professor, Colonel, The Reverend, The Honorable,* etc. Nor should you ever abbreviate a person's name: *Chas., Jas., Wm.*

It is now common practice to address a married woman by her own, rather than her husband's, first name: *Mrs. Marie Jones,* not *Mrs. Robert Jones.* You may also address a woman, married or unmarried, as *Ms. Marie Jones,* unless you know that the person objects to this form (as some do).

Opinion is divided on the use of two-letter state abbreviations in the inside address. Many organizations prefer that the state name be spelled out in the letter itself but abbreviated in the outside address. A sample list of inside addresses, together with appropriate salutations, is included as the third main section of this discussion.

Salutation. A general rule for business correspondence is that the salutation should match the inside address. Accordingly, in a letter addressed to *Mr. John Smith,* the recipient is addressed as *Dear Mr. Smith.* If the address includes some other title—*Dr., Professor,* etc.—the recipient is addressed by that title.

In a letter addressed to an organization (with no particular addressee), the correct salutation is *Gentlemen* (or *Ladies* if that is appropriate). When you are uncertain whether you are addressing women or men, you could use *Ladies and Gentlemen. Dear Sirs,* formerly used as a group salutation, is now considered somewhat pompous and out of date.

Dear Sir, once in vogue in business letters, is also considered too stuffy for all but the most formal letters, such as one to a member of Congress, state governor, or other high official. *Dear Sir or Madam,* however, may be necessary in a letter addressed to a company official whose name is unknown to the writer, e.g., the *director of personnel,* the *advertising manager.* In this situation, a modern trend is to address the individual by title, as, "Dear Personnel Manager" or "Dear Advertising Manager." (See p. 89.)

Body. The body of a letter is the message conveyed. Each paragraph may be indented five spaces or may begin at the left margin (block style). The latter is now the preferred style in most organizations. In all but the briefest letters, single spacing, with an extra space between paragraphs, is used.

Complimentary Close. The complimentary close is placed a double space below the final line of the body and is usually aligned with the typewritten heading (or the date if the heading is printed). The closings most used today are *Very truly yours* and *Sincerely yours*. (Note that only the first word is capitalized.)

Cordially yours is sometimes used but would be inappropriate in certain situations, as in a letter addressed to someone the writer knows only slightly, a letter requesting a favor, or a letter seeking collection of an overdue bill. *Respectfully yours,* which was in vogue some years ago, is now considered too formal except when the recipient's position—a member of Congress or other government official—indicates that the respect is truly intended.

The old participial endings—*hoping this finds you in good health, I remain*—went out with the horse and buggy and should remain forever in oblivion.

Signature. The signature line is typed at least four spaces below the complimentary close to leave room for the writer's signature. If the writer is signing for an organization, his or her position is usually typed just below the name (see samples).

When a woman is signing a business letter, she may wish to indicate her marital status for the convenience of whoever will answer the letter. This can be done by typing *Mrs., Miss,* or *Ms.* (usually in parentheses) just ahead of the typed signature. If no indication is included with the typed signature, her title may be assumed to be *Miss.* A married woman may elect to use her husband's name in the typed signature.

Here are several complimentary closes and typed signatures:

Very truly yours,
Robert L. Smith

Cordially yours,
Janet L. Smith

Sincerely yours,
Robert L. Smith
General Manager

Sincerely yours,
(Mrs.) Janet L. Smith
Credit Manager

Identification Line. The identification line shows the initials of the person who dictated the letter in capitals and the initials of the typist in lower case. In a block style letter, the identification line is typed a double space below the typed signature. The letter on page 93 shows an example of an identification line.

Attention Line. In a letter addressed to a firm, an attention line is sometimes used; the line is placed a double space below the inside address and a double space above the salutation. The attention line indicates the person who should read the letter, even though it is not addressed to him or her personally. The use of an attention line is illustrated below.

Enclosure. The sample letter on p. 99 shows the correct form for indicating material enclosed with a letter.

Forms for Inside Addresses and Salutations

Letters Addressed to Individuals, Groups, Firms, Etc.

An Individual

Mr. Harold Brown
1416 High Street
Springfield, Illinois 62788

Dear Mr. Brown:

Ms. Janet Brown
2440 Sumpter Drive
Lima, Ohio 45810

Dear Ms. Brown:

An Individual in a Firm

Mr. Robert L. Johnson, Manager
American Insurance Corporation
3729 State Street
Cleveland, Ohio 44900

Dear Mr. Johnson:

To a Firm with Attention Line

American Insurance Corporation
3729 State Street
Cleveland, Ohio 44900

Attention of Mr. John H. Brown
Dear Sir:
(or)
Attention of Mrs. Marjorie T. Brown
Madam:

To a Company Official, Name Unknown to the Writer

The Advertising Manager
Johnson and Smith Company

1643 West Front Street
Spokane, Washington 99201

Dear Sir (or) Madam: (or) Dear Advertising Manager:

To Two Persons

Mr. William Bell and Mr. Robert L. Camp
3942 Entrance Street
Fair Oaks, California 95628

Gentlemen:
(or)
Mrs. Carol B. Taylor and Ms. Alice Brown
Western Products, Incorporated
6516 Madison Street
Tulsa, Oklahoma 74175

Ladies:

A Group of Women

Western Women's Association
901 Sixth Street, Suite 204
Denver, Colorado 80216

Ladies:

A Doctor (M.D.)

Dr. Robert G. Brown
(or)
Robert G. Brown, M.D.
2826 High Street
Savannah, Georgia 31412

Dear Dr. Brown:

Letters Addressed to Educators.

President of University or College

Dr. (or) Mr. (or) Ms. ___________
President of ___________ University
Appropriate address

Dear Sir: (or) Madam:
(or)

My dear President ___________:

Other University Officials

Dr. ___________
Dean of Men (or) Dean of Women
___________ University
Appropriate address

Dear Dean ___________:

Doctor of Philosophy (or) Law (or) Medicine (or) Divinity

Harold F. Smith, Ph.D. (or) LL.D. (or) M.D. (or) D.D.
(or)
Dr. John R. Smith (or) Dr. Miriam R. Smith
___________ University
Appropriate address

Dear Dr. Smith:

Chairperson of Division or Department

Dr. ___________
Chairperson, Humanities Division
___________ University
Appropriate address

Dear Dr. ___________:

Professor

Professor ___________
Humanities Division
___________ University
Appropriate address

Dear Professor ___________:
(or if Ph.D.)
Dear Dr. ___________:

Outside Addresses

To facilitate mechanical sorting of envelopes, the United States Postal Service has devised two-letter abbreviations for all states and U.S. territories. The abbreviations, along with the ZIP codes, can be read by a scanner, which enables automatic sorting of letters. The Postal Service encourages the use of state abbreviations and ZIP codes on all letters. If the abbreviation and ZIP code do not appear on the envelope, the scanner will reject the letter, which must then be hand-sorted.

Local ZIP codes are found in most telephone directories. ZIP codes for any area may be obtained from any post office. Following are the two-letter abbreviations for the fifty states and the District of Columbia, the Canal Zone, Guam, Puerto Rico, and the Virgin Islands:

Alabama	AL
Alaska	AK
Arizona	AZ
Arkansas	AR
California	CA
Canal Zone	CZ
Colorado	CO
Connecticut	CT
Delaware	DE
District of Columbia	DC
Florida	FL
Georgia	GA
Guam	GU
Hawaii	HI
Idaho	ID
Illinois	IL
Indiana	IN
Iowa	IA
Kansas	KS
Kentucky	KY
Louisiana	LA
Maine	ME
Maryland	MD
Massachusetts	MA
Michigan	MI
Minnesota	MN
Mississippi	MS
Missouri	MO
Montana	MT
Nebraska	NE
Nevada	NV
New Hampshire	NH
New Jersey	NJ
New Mexico	NM
New York	NY
North Carolina	NC
North Dakota	ND
Ohio	OH
Oklahoma	OK
Oregon	OR
Pennslyvania	PA
Puerto Rico	PR
Rhode Island	RI
South Carolina	SC
South Dakota	SD
Tennessee	TN
Texas	TX
Utah	UT
Vermont	VT
Virginia	VA
Virgin Islands	VI
Washington	WA
West Virginia	WV
Wisconsin	WI
Wyoming	WY

Sample Letters

Requesting a Favor. When asking a favor, a writer should try to offer something in return. If the recipient sees a possible benefit in granting the request, he or she will probably answer at once. The following letter was sent by a public agency requesting information and photographs for inclusion in a brochure that was to explain the reclamation of waste water. The letter makes clear that by complying with the request, the recipient will receive wide publicity. (Letters from public agencies or business organizations would of course be on letterhead paper.)

Letter of Request (showing block style)

June 15, 1981

Mr. Edward Smith, Manager
ABC Reclamation District
Your Town, Arizona 85941

Dear Mr. Smith:

We are planning to publish, for release to the public, a brochure showing the latest techniques for the reclamation and reuse of waste water. We particularly want to point out and illustrate with diagrams and photographs: (1) treatment facilities, (2) reclamation plants, and (3) uses for reclaimed water.

Since your facility was one of the first in Arizona to treat and use reclaimed water, I believe it will illustrate what we want to emphasize. I would appreciate your sending me information and photographs covering those three general aspects, and, in particular, the use of reclaimed water within your district.

We hope to make ABC a prominent part of our brochure.

Very truly yours,

Robert Thompson

Robert Thompson
General Manager

RT: eh

Requesting a Recommendation. A letter asking for a personal recommendation would naturally be sent to someone who knows the writer well. In this letter, a recent college graduate is asking a former instructor to send a letter of special recommendation:

Request for a Recommendation (showing modified block style)

1637 Norwood Way
Center City, California 90002
October 15, 1981

Professor Harold P. Wiley
Department of Journalism
California State University, Sacramento
6000 J Street
Sacramento, California 95819

Dear Professor Wiley:

I am delighted to tell you that I am being considered for promotion here at Radio Station KXXX. The position of assistant news director is open, and I have made a strong bid for the job. Since I was your news director during my senior year, I thought you might be willing to send along a letter explaining what I did under your direction.

Because the management plans to decide on the position in about ten days, I would appreciate your sending a letter immediately. You may write to:

Mr. Donald Gordon, Manager
Radio Station KXXX
3814 High Street
Center City, CA 90003

I still remember all the fun and excitement at KCCC, but I am naturally happy to be started in commercial broadcasting. I trust the old college transmitter is still holding together.

Warmest personal regards.

Yours very truly,

James C. Horner

James C. Horner

Letter of Complaint. A letter of complaint or one requesting an adjustment is likely to be poorly written, for if the writer is emotionally upset, he or she may take out his or her frustrations on the recipient. In most cases, however, an angry or sarcastic letter will do little more than antagonize the reader and thus get the writer nowhere.

When writing a letter of complaint, always be guided by the objective you hope to attain—perhaps a refund, return of defective merchandise, or adjustment of a bill. Even if your momentary desire to tell off the reader were justified, you might well defeat your main purpose by doing so. Therefore a letter of complaint should be completely courteous, although it need not be weak or apologetic. In most cases, a straightforward recitation of the facts will get the reader's attention. And, if you state the facts accurately and logically, you may well have the recipient agreeing with you before he or she reaches your request for an adjustment or refund.

Here are two such letters, the first requesting an adjustment, the second inquiring whether a repair bill is accurate:

Letter of Complaint Requesting an Adjustment

5541 Westminster Drive
Central City, Texas 75064

July 20, 1981

Mr. John Hammond, President
Central Motors Company
3310 Beverly Drive
Central City, Texas 75063

Dear Mr. Hammond:

Since we have just returned from vacation, this is the first chance I've had to tell you about the used car we bought at Central Motors on Saturday, June 28. We actually purchased the car the night before; after we had closed the deal, however, the salesman discovered that the car needed an oil change and lubrication, and he agreed to have it ready by Saturday noon.

We bought the car to go on vacation, and, since we had planned to leave at noon on Saturday, everything seemed to be on schedule. However, when we arrived to pick up the car, it wasn't ready. Instead, the salesman asked us to come back on Monday; since it was a Saturday, he had been unable to have the car serviced, and your service area was closed until Monday. That arrangement, of course, was impossible.

As calmly as possible, I took the "new" car, went to our regular service station, spent $22 for servicing, and didn't get away until 4 o'clock. Despite what that did to our schedule, we accepted the delay quite calmly, but I am not so calm about the $22. I realize the salesman was busy and had forgotten (at the time) that the next day was Saturday. However, since we had told him of our plans and he had agreed to have the car ready, I feel it only fair that the $22 be refunded.

Letter of Complaint (cont.)

This may seem like a small item now, Mr. Hammond, but at the time it was frustrating, unsettling, time consuming, and expensive. And since everything was supposedly arranged in advance, I believe the car should have been ready at the agreed-on time. I hope you can appreciate my frustration. I have enclosed the service bill for your consideration.

Very truly yours,

Donald G. Rogers

Enclosure

P.S. The car was everything the salesman said it was, and our vacation trip was trouble-free.

Letter Questioning a Repair Bill

4504 High Street
Hastings, Nebraska 68901
August 15, 1981

Johnson Hardware Company
6445 Main Street
Hastings, Nebraska 68901

Dear Sir:

Yesterday your man was here to install a new switch on my dishwasher. The machine works fine now, but it seems to me that $37.75 is a high price for such a little item. And I never did understand what the serviceman did. I was babysitting my grandson at the time, and in all the confusion, your man left without giving me a breakdown of the costs. Are you certain that $37.75 is correct? If he made a mistake, I would like a refund.

Very truly yours,

Mrs. Mary M. Williams

Answering a Letter of Request. If you plan to grant a request, say so immediately (good news first). On the other hand, if you must refuse, begin by giving logical reasons for your refusal. If your reasons are sound, the recipient may well be agreeing with them before he or she gets to the refusal. Then, refuse the request as positively as possible. When the request is for information, suggest another source, if you know of one where the request will likely be honored. Then, close positively without repeating the apology for your refusal. Here is an answer to the letter of request on p. 93.

June 22, 1981

Mr. Robert Thompson
General Manager
XYZ Water Agency
Midtown, Arizona 86045

Dear Mr. Thompson:

Thank you for your invitation to be included in your new brochure on water reclamation. We appreciate this opportunity to show the people of Arizona what we are doing.

At the moment, however, we are in the midst of an overhaul, so to speak. We are converting our advanced treatment plant to ion exchange, and any information we supplied now would be somewhat premature. If we give you information on our former processes, your brochure would be out of date, at least in our case, before you published it. I am certain you can appreciate this dilemma.

Have you written to Midland Water District for information? Charles Durand, their general manager, is proud of their new secondary operation, and I know he would be pleased to have the plant included in your brochure.

If you plan a further publication, please try us again in about six months. We'll be back in full operation by then, and we'll be glad to help out when we are back to normal.

Good luck on your forthcoming brochure. And please include us on your mailing list.

Cordially,

Edward Smith

Edward Smith
Manager

Answering a Complaint or Request for Refund. A letter of explanation need be neither loquacious nor humble. In such a letter you should apologize briefly, explain the circumstances that caused the complaint, and, as quickly as possible, switch to a positive tone. If you intend to grant an adjustment or refund, you have no problem. Begin by telling the recipient he or she is right and you were wrong, and that a refund or adjustment is in order (good news first). Then close the letter on a positive note without repeating the apology or the cause of the complaint.

The letter refusing an adjustment, however, requires more tact. In this sort of letter, as in a letter refusing a request, you should begin by stating logical reasons for your refusal. As in a letter of complaint, if your reasons are accurate and logical, the reader may begin to see your side of the controversy before getting to the refusal. Yet, even when you are 100 percent right and the recipient is dead wrong, don't jump in with both feet. In a dispute over a service charge, for instance, never prove that a reader cannot add or subtract. This is really an extension of the "you attitude" (p. 85), which all letter writers should adopt. If you keep the "you attitude" in mind, you will remember to suggest that you have considered both sides of a controversy—the reader's side and your own.

After reciting the facts leading to your refusal, you should avoid such expressions as "Therefore we must refuse," "We cannot accept," and other negative expressions. Instead, try some more tactful approach. Note the difference in these two statements:

Blunt: We must refuse your request for a refund.
Tactful: In view of the facts, our charge seems correct.

Here are two letters of apology. The first is a reply to the unhappy car buyer; the second explains the charge for the dishwasher repair and (indirectly) refuses an adjustment.

Letter Granting a Refund

July 20, 1981

Mr. Donald G. Rogers
5541 Westminster Drive
Central City, Texas 75064

Dear Mr. Rogers:

You're so right about that $22, and I am enclosing our check for that amount with this letter. I certainly can appreciate your feelings about the delay and the extra expense, and I want you to know that at Central Motors we do try to stand behind our agreements. Thank you for telling me what happened.

Thank you, also, for letting us know that the car performed so well on your trip. And please remember to come in at the end of thirty days for the free checkup that goes with every used car purchased here. We want to be sure that yours continues to give you trouble-free performance. We'll be looking for you about August 1.

Cordially,

John Hammond
President

JH: at

Enclosure

Letter Explaining a Service Charge

August 17, 1981

Mrs. Mary M. Williams
4504 High Street
Hastings, Nebraska 68901

Dear Mrs. Williams:

I have your letter of August 15, and you are absolutely right. Mr. Johnson, our service representative, did not leave you a copy of the invoice explaining the service charge of $37.75. Here is a breakdown of the charges:

Although the switch was just $13, Mr. Johnson also installed a new wiring harness. The harness was $7.50, and the sales tax on both items was $1.25. To this we added $16 to cover the expense of the service call. Accordingly, our charge seems correct. I have enclosed your copy of the invoice with all the costs broken down.

The installation is of course guaranteed for one year. Mr. Johnson also inspected the other components of your dishwasher, and, according to his report, it should continue to give you excellent service.

Letter Explaining a Service Charge (cont.)

```
Thank you for calling Johnson Hardware.  Please try us again.

Very truly yours,

John L. Jones
Store Manager

JLJ: ap

Enclosure
```

Those two examples illustrate the "formula" for letters of apology, letters granting adjustments, and letters refusing adjustments. Remember these principles:

- Apologize briefly but don't overdo it. Avoid restating the cause of the complaint.
- If you are granting an adjustment, say "yes" immediately.
- If you are refusing an adjustment, first give logical reasons for your refusal. Then, say "no" positively and tactfully.
- End your letter on a positive note. Don't write a "sorry" closing.

See also **Application Letters and Résumés.**

COUNCIL. See **Consul, Council, Counsel.**

CRITERIA, CRITERION. Remember that *criteria* is the plural form of *criterion* and requires a plural verb, adjective, or pronoun. For example:

1. The test criteria *are* (not *is*) as follows: . . .
2. The discussion is based on *these* (not *this*) criteria.
3. Each *criterion* was carefully tested.

CROSS (as a prefix). See **Prefixes and Suffixes.**

D

DAMAGE, DAMAGES. The singular, *damage,* is a collective noun that means the total harm resulting from an accident, fire, earthquake, etc. The plural form, *damages,* is a legal term denoting monetary compensation for damage or injury. If you were responsible for *damage* to a neighbor's fence, you might have to pay the *damages*.

DANGLING CONSTRUCTIONS. Dangling constructions, which relate to words they cannot logically modify, can not only embarrass writers but also mislead readers. The most common are the dangling participle and the dangling infinitive.

Dangling Participle

Relaxing in a warm bath, the telephone rang three times.

Analysis: The participial phrase (relaxing in a warm bath) modifies the subject, *telephone,* which it cannot logically do. Note that the participial phrase has no subject and thus must logically modify the subject of the main clause. *Solution:* Make certain the opening phrase logically modifies the subject.

Revision

While I was relaxing in a warm bath, the telephone rang three times.

Here is an example of a misleading "dangler":

Dangling Participle

Coming out of the dark cave, the air felt fresh and cool.

Analysis: The participial phrase (coming out of the dark cave) has no subject. Its function is to modify the subject of the main clause, which results in a misleading statement. The writer meant the fresh air outside the cave; his construction has the air coming out of the cave.

Revision

Coming out of the dark cave, we felt the fresh, cool air.

Now the statement is logical, but a better construction might be:

As we came out of the dark cave, the air seemed fresh and cool.

Dangling Infinitive

To get the most out of the course, the book must be studied carefully.

Analysis: The infinitive phrase (to get the most . . .) modifies *book,* which it cannot logically do.

Revision

To get the most out of the course, you must study the book carefully.

See also *Be Careful with Modifiers*, part 1, p. 20.

DASH. The dash, which functions somewhat like a **comma,** is used to indicate an abrupt change in sentence structure. Use it:

1. To set off a group of words, each of which is separated by commas:

   Every use of water—in homes, in factories, on farms—results in a change of water quality.

2. To separate a summarizing expression that follows a list:

   The knife sharpener, the grindstone, the floor buffer—all of these accessories are very handy.

3. To set off a parenthetical thought:

   Much of the extra cost—possibly 90 percent of it—was absorbed by the contractor.

4. To set off a thought or phrase for special emphasis:

   I want just one thing after graduation—a good job.

Note: Dashes should be used sparingly and only to indicate a break in normal sentence construction. If a dash is loosely substituted for a comma, it will lose some of its effectiveness as a distinctive mark.

A dash is indicated on the typewriter with two hyphens.

DATA. Data, the plural form of the Latin *datum,* is a **collective noun** denoting facts and figures from which an inference may be drawn. The word has inspired endless argument about whether a singular or plural verb should be used with it—data *is* or data *are.* Although the form is plural, many writers use data *is,* especially to refer to a group of data from the same analysis, test, event, etc. According to *Webster's Third New International Dictionary* (unabridged), either usage is acceptable.

DATES. To express dates, simply use the number designating the day of the month without *st, th, rd,* etc.; for example, May 9, 1980 or 9 May 1980. Dates are sometimes written out in formal announcements and invitations: the fifteenth of May, nineteen hundred and eighty.

DEADWOOD. Deadwood refers to the extra words that add nothing to the meaning of sentences. For example:

Inflated: The high-pressure, superheated steam actuates the turbine, which *serves to* turn the generator rotor and produce electrical energy.
Concise: The high-pressure, superheated steam actuates the turbine, which turns the generator rotor and produces electrical energy.

Inflated: For the *past* two years, the company has been *engaged in a program for the recruitment of* electrical engineers.
Concise: For two years, the company has been actively recruiting electrical engineers.

For other examples of how words are wasted, see *Write Concisely (But Not Too Concisely)*, part 1, p. 17 and **Inflated Expressions** (Handbook).

DEFECTIVE, DEFICIENT. Each of these words indicates that something is lacking. *Defective* refers to quality, whereas *deficient* refers to quantity. For example, a machine that does not operate properly is *defective;* a shortage of machines is a *deficiency*.

DEFINITE, DEFINITIVE. Both words mean precise, clear, and certain. *Definitive,* however, means "not subject to change." For instance, a plan may be *definite* because it has been worked out in full detail but *definitive* only after it is no longer subject to amendment.

Definitive is also used to designate a work considered the most authoritative on its subject. For example: "For sixty-three years, Einstein's theory has been considered the *definitive* work on relativity."

DEFINITIONS. Technical writers frequently use terms that must be defined, sometimes in a single sentence but often in a longer (expanded) definition that may extend to several sentences or even to several paragraphs. The importance of clear definitions cannot be overemphasized. In a sentence definition, for example, it is usually risky to slip in a short dictionary definition, which may be only a synonym or a "circular" definition. An example of the former: "A permutation is a transformation"; of the latter: "permeable means capable of being permeated." As you can imagine, such brief definitions would leave most readers talking to themselves.

Sentence Definitions

A clear sentence definition requires two parts, often called the *genus* (the class) and the *differentia,* which "differentiates" between members of the same class. If you remember to include both parts in a simple definition, you will find it easier to make yourself understood. Reduced to a formula, a definition (D) becomes the genus (g) plus the differentia (d). Thus, $D = g + d$.

To define *crucible,* for example, you would find the following short dictionary definition inadequate: *a vessel of refractory material.* The term *refractory* creates its own mystery, and there is no hint of how the "vessel" differs from other vessels. Applying the formula $D = g + d$ should help you create a clear definition of the term:

A crucible =
Genus: a vessel of heat-resistant material, such as porcelain or graphite +
Differentia: used for melting minerals and metals.

Thus the formula makes possible a general definition that would give most readers an idea of what the term entails. That definition would be useful in a sentence like this:

The gold is then placed in a crucible, a vessel of heat-resistant material—such as porcelain or graphite—used for melting metals and minerals.

In that context, perhaps an explanation of how gold inlays are made in a dental laboratory, the preceding simple definition would be adequate.

If, however, your purpose were to define crucible more completely, you would have to include more information, for instance, the distinction between a crucible used in a dental laboratory and one used in the production of high-quality crucible steel. You might also point out that electric-arc and induction furnaces have largely supplanted the crucible in modern steel production and that use of the crucible today is chiefly confined to the manufacture of tool steel. With such additional information you could expand, or extend, the definition to present a clearer, more concrete picture.

Expanded Definitions

An expanded definition often follows a pattern of comparison and contrast, in which a writer compares or contrasts an item or concept with a similar or dissimilar item or concept. Writers may also clarify definitions by providing specific examples of a term or by explaining something the term seems to include but does not. Still another useful method is to divide a term into parts, or components, to help clarify an abstract idea. Often these patterns will overlap; that is, a writer may use more than one in the same definition.

Specific examples. A specific example will help clarify a vague term that may be difficult for readers to visualize. In the following brief definition of recycled water, the writer lets the readers see an example:

Recycled water is water that is directly reused without treatment, at the same general location and usually for the same purpose. For example, recycling takes place when excess irrigation water is drained from a field, mixed with incoming water supplies, and reused on that field or an adjoining field. In a typical farm reuse system, small sumps at the lower end of an irrigated field collect surplus applied water; the water is then pumped to a "head ditch," where it is mixed with the incoming water and used again.

Comparison and Contrast. In the following definition of two statistical terms (*population* and *sample*), the writer provides examples and uses comparison and contrast to bring out their exact (statistical) meanings:

A statistician may look at a set of data (records) as a population or as a sample. For instance, if the records contained all possible observations of a certain event, they would make up a population; if they contained only part of the possible observations, they would be called a sample. Whether the records are considered a population or a sample also depends on how they will be used.

More specifically, if a statistician had complete records of the academic averages of all graduates of a certain high school in a particular year, she would call the data a population, provided she were interested in those students only. If, however, she intended to use the records as a basis for generalizations about all high school graduates in the northern part of the state, she would call the data a sample.

See also **Comparison and Contrast.**

Division into Parts. In this brief definition, the writer divided the abstract term *biology* into two specific parts:

The science of biology includes the origin, growth, reproduction, structure, and life cycle of all forms of life. Biology can be broadly divided into botany, or the science of plants, and zoology, the science of animals.

Scientific Definitions

Technical and scientific writers frequently have to clarify scientific terms for lay readers. In the following expanded definition, the writer used comparison and contrast, a liberal number of examples, and considerable imagination to explain some distinctive properties of water:

> Of all the natural substances on earth, only water can be found in three distinct forms at the same time—as a solid, a liquid, or a gas. For instance, on a winter day, water appears as ice in a frozen creek, as a liquid in a flowing river, and as clouds in the sky. Water expands when it freezes whereas most other substances contract. Most substances are heavier in solid form than in liquid form; yet a block of ice is lighter than an equal volume of water.
>
> Water takes its time to boil—a vexing fact to those in a hurry for their morning coffee. That the watched pot boils slowly is a humorous but nonscientific attempt to explain another distinctive property of water—its ability to absorb a great amount of heat without much rise in temperature. In fact, water's heat capacity is so high it is used as a standard against which the heat capacities of other substances are measured. Because of this ability to absorb heat, the oceans act as controls in preventing extremes of climate.
>
> Most remarkable of all of water's unusual properties is its ability to dissolve so many other substances. Of all known liquids, water is the nearest to a universal solvent. If table salt, for instance, is added to a glass of tap water, the salt crystals will dissolve and become equally distributed as ions, or electrically charged atoms. When table salt dissolves in water, it will be distributed as positively charged sodium ions and negatively charged chloride ions. In this instance, the water is the solvent, the salt is the solute, and the resulting homogeneous mixture is called the solution.[7]

DEMONSTRATIVE PRONOUNS. See **Pronoun.**

DENOTATION. See **Connotation and Denotation.**

DEPENDENT CLAUSE. See **Clauses and Phrases.**

DESCRIPTION. A description—of a house, a scenic area, a community, etc.—usually follows a spatial pattern, for example, from left to right, top to bottom, north to south. Sometimes a description centers on a particular

[7]Adapted from U.S. Geological Survey pamphlet, *A Primer on Water Quality* (Washington, D.C.: Government Printing Office, 1965).

feature and proceeds in a "circular" pattern. Writers should be aware of two important facts about descriptions: (1) a description cries out for supporting details; (2) the details must be specific, not vague generalizations. Accordingly, writers should avoid such vague terms as *colorful, lofty, spacious,* in favor of words that let readers envision the colors, the loftiness, or the spaciousness.

Technical writers often have to prepare descriptions of buildings, equipment, machinery, motors, etc. Usually, a writer prepares some type of illustration(s) to show readers what is being discussed. Sample pages showing the use of photographs, charts, and other illustrations are presented under **Graphic Aids** (pp. 136–149). The illustrations have been placed in that section so that several types of graphic aids can be presented as one unit.

In the following description of a dam and powerhouse, the writer follows a spatial pattern and provides exact dimensions to help readers visualize the size of the structures:

> Oroville Dam, on the Feather River in northern California, is the highest earthfill dam in the United States. It rises 770 feet above the river bed and spans 5,600 feet between abutments at its crest. The dam embankment consists of 80 million cubic yards of impervious clay resting on a concrete core block.
>
> The spillway, located on the right abutment of the dam, comprises two separate elements: a gated flood-control outlet and an uncontrolled emergency spillway. The flood-control outlet consists of an unlined approach channel, a gated headworks, and a lined chute extending to the river. The emergency spillway is a concrete overpour section 1,730 feet long.
>
> Most of the water released from the reservoir passes through a power plant located in the left abutment. The power plant intake is a sloping concrete structure consisting of two parallel intake structures, one for each of the 22-foot-diameter penstock tunnels, which are protected by stainless-steel trash racks at the intake openings.
>
> The powerhouse, located underground, is 550 feet long, 69 feet wide, and 140 feet high. The plant's 678-megawatt output is derived from three conventional generators.

Description of a Mechanism

When describing a mechanism, a writer must be careful to include all the parts that make up the item under discussion. If the description will be widely read, the writer must avoid using language that only experts will understand. Here is part of one writer's description of the automobile ignition system.[8]

The function of the ignition system is to ignite the combustible mixture in each engine cylinder at the instant that will yield maximum power and efficiency from the explosion. Ignition is accomplished by a high-tension spark that jumps a gap between two electrodes in the combustion chamber. The electrodes are contained in a spark plug consisting of two parts: a center electrode and an outer steel shell held together by a removable bushing (metal lining) or, more commonly, sealed at the time of manufacture.

The center steel shell, threaded into the cylinder head, is grounded to the engine block; the center electrode, consisting of a steel wire surrounded by insulating ceramic, is connected to the high-tension source. The spark gap varies from 0.025 to 0.040 inches (0.6 to 1 millimeter), measured from the face of the center-electrode wire to a wire extending horizontally from the steel shell.

The spark plug must be able to withstand high pressure without leaking, and the insulation between shell and electrode must be of high insulating quality, able to withstand high temperatures and shock loads, and be a good conductor of heat. If heat is not conducted away from the electrodes, preignition may result.

The source of high-tension current is a coil sealed in a cylindrical metal case. The coil consists of primary and secondary windings wrapped around a central core of soft iron. When current passes through the primary winding from the system's 12-volt battery, a strong magnetic field forms around the windings. Every time the current flow in the primary winding is interrupted, the rapid collapse of the magnetic field induces an electrical potential of up to 20,000 volts in the secondary winding, sufficient to cause a spark to jump between the electrodes of the spark plug.

DIACRITICAL MARKS. Diacritical (accent) marks are phonetic symbols used to indicate pronunciation. They include the phonetic symbols used in many dictionaries and those used to indicate the pronunciation of certain foreign words (see **Foreign Words**). Diacritical marks are seldom used in ordinary writing, however, as they are not found on the standard typewriter keyboard. Here are the most common diacritical marks:

[8]Reprinted with permission of *The Encyclopedia Americana*, copyright 1978, The Americana Corporation.

Name	*Symbol*	*Example*	*Indicates*
acute	´	attaché	stress on the letter marked and that the *e* is pronounced like a "long" *a*
breve	˘	săck	a "short" vowel sound
cedilla	¸	façade	an *s* sound when the mark is placed beneath a *c*
circumflex	ˆ	crêpe	a muted vowel sound
dieresis*	¨	naïve coördinate	that two consecutive vowels are pronounced separately
grave	`	père	a deep vowel sound articulated at the back of the mouth
macron	¯	essāy	the "long" vowel sound
tilde	˜	piñon	the Spanish *ny* sound as in *canyon*

*The dieresis should not be confused with the umlaut (m̈), which is placed above *a, o,* or *u* in some German words to indicate that the vowel sound is divided: ä = *ae;* ö = *oe;* ü = *ue.*

DIAGNOSIS, PROGNOSIS. The distinction between these two clinical-sounding words is often overlooked. Keep in mind that your doctor's *diagnosis* is his or her identification of your medical problem. The *prognosis* is the doctor's estimate of your prospects for recovery. If, for instance, the doctor diagnosed your problem as influenza, the prognosis might be that, with proper care and rest, you would be back to normal in about ten days.

DICTION. Diction, meaning a writer's choice and use of words, is critical to the effectiveness of all writing. See *Choose Words Carefully*, part 1, p. 14, and **Colloquial Language; Homonyms; Inflated Expressions; Malapropisms; Pompous Language; Sexist Language; Spelling; Technical Terms and Jargon.**

DIFFERENT. *Different* is sometimes used redundantly or as deadwood. For example:

1. I received estimates from three *different* contractors.
2. I visited San Francisco on three *different* occasions during 1980.

Note: Different is implied by the word *three;* thus the word itself is superfluous. The use of *different* in the second example results in a rather pompous sentence, which means simply, "I visited San Francisco three times during 1980."

DIFFERENT FROM, DIFFERENT THAN. Although opinion is divided on which form is acceptable, *different from* is preferred by most authorities:

1. His opinion is *different from* mine.
2. The house was quite *different from* the description I had read.

On the other hand, different *than* is preferable when the use of *from* would result in awkwardness or verbosity:

1. The test results were considerably different *than* I had expected (instead of *from what* I had expected).
2. The weather today was quite different *than* yesterday (instead of *from what it was* yesterday).

DILEMMA, PROBLEM. Strictly speaking, these two words are not synonymous. A *dilemma* denotes a situation offering a choice between two alternatives. For instance, a college senior might be offered a promising job with an excellent starting salary. Against this prospect he must weigh the fact that he has been planning to return to school for an advanced degree, which may offer even greater future opportunities. This graduate is faced with a *dilemma:* he must choose between two equally attractive opportunities.

On the other hand, if a motorist runs out of gas ten miles from the nearest service station, she has a real *problem.*

DIRECT OBJECT. A noun or noun equivalent that receives the action of a transitive verb is called the direct object. It may be a noun, pronoun, gerund, infinitive, or clause:

Noun: John bought a *car* last night.

Pronoun: I took *her* to the dance.

Gerund: Mary loves *dancing.*

Infinitive: I love *to dance.*

Clause: He took *what he wanted.*

See also **Indirect Object; Verb.**

DISCREET, DISCRETE. To be *discreet* is to act with discretion, i.e., to show good judgment, to be tactful, or to show respect for accepted standards of propriety.

Discrete means separate or individual, as consisting of many individual, discrete parts. It is particularly applicable to mathematical expressions, referring to individual values.

DIVISION OF WORDS. The division of words at the end of lines is really a typing or printing problem. However, writers should know how to divide words and when not to divide them.

1. Do not divide:
   a. Numbers or numerical designations, e.g., *Section II, 35 inches, page 10.*
   b. Words of one syllable, such as *though, through, knock, guard, home, ground.*
   c. If the division would be misleading as to the meaning or pronunciation. For example, do not divide such words as *often, motor, noisy, water.*
   d. If the result would leave a single letter on one line. Do not divide such words as *again, among, enough, event, item, unite.*
   e. The following suffixes:

ceous	cious	gious	tial
cial	geous	sial	tion
cion	gion	sion	tious

   f. Expressions containing hyphenated words.
   g. Proper nouns, especially names of persons.
2. Divide words according to the following rules:
   a. Divide according to pronunciation.
      knowl-edge NOT know-ledge
      perform-ance NOT perfor-mance
      injec-tor NOT inject-or
      radia-tion NOT radi-ation
   b. With certain exceptions, divide words after a vowel unless this violates rule a.

physi-cal NOT phys-ical
particu-lar NOT partic-ular
sepa-rate NOT sep-arate
fila-ment NOT fil-ament

Exception: When the divided word ends in *able* or *ible,* the vowel is carried over to the next line:

convert-ible, cast-able, read-able, flex-ible.

c. When the pronunciation warrants it, separate two consonants standing between vowels.

advan-tage	exces-sive	impor-tant
pos-sible	propel-lant	struc-ture

d. With certain exceptions, carry over the *ing* in present participles.

alter-ing	certify-ing	chang-ing
cast-ing	whirl-ing	revok-ing
enter-ing	improvis-ing	infiltrat-ing
seek-ing	investigat-ing	check-ing

Exceptions: When the ending consonant is doubled before the addition of *ing,* the added consonant is carried over.

control-ling	plan-ning	bid-ding
transfer-ring	run-ning	grip-ping

When the final consonant sound belongs to a syllable with a silent vowel, the final consonants become part of the added syllable *ing.*

dwindle	baffle	handle
dwin-dling	baf-fling	han-dling

DOCUMENTATION. In certain types of exposition, some form of documentation may be necessary to acknowledge sources on which the information is based. This is particularly true of reports and other work based on scientific research. Documentation is most often used to (1) cite authority for certain statements, (2) acknowledge the source of a direct quotation, and (3) tell readers where additional information may be found.

The conventional methods of documentation are (1) footnotes; (2) a numbered list of references, each of which has been cited by number in the text; and (3) a complete bibliography at the end of the text. Sometimes both footnotes and a bibliography are provided.

1. *Footnotes*

A footnote is indicated in the text with a number or an asterisk raised one-half space above the appropriate line. For example:

The estimated annual cost is $100,000.[1]

Note that the number following the footnote follows all other punctuation (except a dash) without a space. If footnotes are numerous, and particularly when more than one appear on a page, it is simpler to indicate them with consecutive numbers than with asterisks, double asterisks, etc.

For research papers and dissertations, most schools require the forms prescribed by the Modern Language Association (MLA). In MLA style, footnotes are typed double space, with a double space between each, and are separated from the last line of text with a quadruple space. The footnote number is indented five spaces, and succeeding lines return to the margin. For more information on MLA style, refer to the latest *MLA Handbook* (published by the Modern Language Association in New York).

This book follows the style prescribed for the natural sciences by the University of Chicago *Manual of Style.*[9] The footnote number begins at the margin and succeeding lines return to the margin. The footnotes are typed single space with a double space between them. Most technical and scientific journals, and most business and governmental organizations, conform to the Chicago *Manual.* The sample footnotes shown here (and the sample bibliography on p. 118) are in Chicago style. (Be sure to follow the style prescribed by your instructor.)

Too many footnotes at the bottom of each page may distract a reader's attention. Therefore, writers may prefer to list them at the end of each chapter (in which case they are called simply notes), or to furnish a **numbered list of references** (p. 117).

See also **Underlining.**

a. *Forms for footnotes*

One-author book

[1]Thomas P. Johnson, Analytical Writing (New York: Harper & Row, 1966), p. 39.

Two-author book

[2]Robert B. Smith and Michael A. Jones, Communicating Technical Information (New York: McGraw-Hill, 1980), p. 86.

[9]*A Manual of Style*, 12th ed. (Chicago: University of Chicago Press, 1969).

No author shown

[3]Sunset Western Garden Book, 4th ed. (Menlo Park, Calif.: Lane Magazine and Book Co., 1979), pp. 41-46.

Article in a book

[4]John D. Roberts, "Management by Committee," in Readings for Managers, ed. Thomas P. Bancroft (New York: McGraw-Hill, 1973), p. 25.

Story or article in an anthology

[5]Lewis Thomas, "Why Can't Computers Be More Like Us?" in Elements of the Essay, selected by H. Wendall Smith (Belmont, Calif.: Wadsworth Publishing Co., 1979), p. 366.

Story, article, or essay in a collection

[6]Andrew Carnegie, "Wealth," in Democracy and the Gospel of Wealth, ed. Gail Kennedy (Lexington, Mass.: D. C. Heath and Co., 1949), p. 3.

Journal article

[7]Willard Thomas, "Use of Video in Technical Training," Technical Communication, 16, No. 4 (1979), pp. 4-5.

Magazine article (signed)

[8]Mark Kramer, "The Ruination of the Tomato," Atlantic, January 1980, p. 72.

Magazine article (unsigned)

[9]"The Great Sell-Off," Time, 14 January 1980, p. 58.

Newspaper story (signed)

[10]Bruce Grant, "Saab Puts 5-Speed in Turbo," Sacramento Bee, 18 January 1980.

(*Note*: According to the Chicago *Manual*, page numbers are unnecessary except in references to the *New York Times* and the *Times* [London].)

Newspaper story (unsigned)

[11]"Simulated Space Shuttle Flight Ended," Sacramento Bee, 18 January 1980.

Government report or article

[12]U.S. Bureau of the Census, Census of Housing: 1978 (Washington, D.C.: U.S. Government Printing Office, 1979), p. 59.

[13]State of California, Dept. of Water Resources, Bulletin 118-80, Ground Water Basins in California (Sacramento: Office of State Printing, January 1980), p. 17.

Encyclopedia article (signed)

[14]George Switzer, "Cryolite," Encyclopedia Americana, 1978 ed.

Encyclopedia article (unsigned)

[15]"Methanol," Encyclopedia Americana, 1978 ed.

Pamphlet (author given)

[16]Robert S. Ayers and Raymond Coppock, Salt Management: California's Most Complex Water Quality Problem (Berkeley, Calif.: U. of Calif., Div. of Agricultural Sciences, 1979), p. 13.

Pamphlet (author not given)

[17]Soil and Water Management for Home Gardeners, Leaflet 2258 (Berkeley, Calif.: U. of Calif., Div. of Agricultural Sciences, 1979), p. 13.

Unpublished thesis

[18]Thomas L. Goodwin, "Removal of Nitrogen from Waste Water by a Symbiotic Process," Diss., U. of Calif., Davis, 1980, p. 13.

Speech or address

[19]G. L. Laverty, "Leaks Make Lakes," Water Conservation Planning Conference, Los Angeles, Calif., 18 January 1980.

Conference proceedings

[20]Proceedings of Second National Solar Radiation Workshop (Huntsville, Ala.: U. of Ala., 1980), p. 76.

Speech from conference proceedings

[21]Frederick Koomanoff, "Monitoring the Solar Resource," Proceedings of the ISES Winnipeg Conference, 1976 (Victoria, Australia: National Science Center), p. 76.

Data manual

[22]Paul Berdahl et al., Solar Data Manual (Berkeley, Calif.: U. of Calif., Lawrence Berkeley Laboratory, 1978), p. 43.
[23]Smithsonian Meteorological Tables, 6th rev. ed. (Washington, D.C.: Smithsonian Institution, 1951), p. 3.

Personal interview

[24]Interview with Donald E. Owen, Chief, Division of Planning, California Department of Water Resources, Sacramento, 10 May 1980.

b. *Abbreviations in footnotes*

(1) *Ibid.* Ibid., an abbreviation for the Latin *ibidem* (meaning *in the same place*), has long been used to indicate that a reference is the same as the immediately preceding reference. If the reference is to a different page, ibid. is used with the appropriate page number. For instance: "Ibid., p. 36." The trend today, however, is to replace ibid. with the author's last name: "Jones, p. 39."

(2) *Op. cit.* Op. cit., which stands for the Latin *opere citato* (in the work cited), is used with the author's last name when a reference has been cited previously but other references have intervened. For example: "Jones, op. cit., p. 40." Op. cit. cannot be used, of course, if the writer has cited more than one book by the same author. Op. cit. too is being dropped in favor of the author's last name: "Jones, p. 40."

Although they are slowly becoming extinct, both ibid. and op. cit. will still be found in many footnotes. They are no longer commonly italicized (or underlined on the typewriter), however.

(3) *Other Abbreviations.* Other abbreviations commonly found in footnotes and other documentation include:

Abbreviation	**Meaning**
c. or ca.	about (Latin *circa*)
cf.	compare with (Latin *confer*)
ch.	chapter
ed.	edition or edited by
et al.	and others (John Smith et al.)
f. or ff.	and the page(s) following
l. or ll.	line(s)
ms or mss	manuscript(s)
N.B.	note well (Latin *nota bene*)
n.d.	no date given
n.p.	no place of publication given

p. or pp.	page(s)
passim	here and there throughout the work
q.v.	which see (Latin *quod vide*)
rev.	revised
tr. or trans.	translated by
vol.	volume

D

In addition to its conventional use for documentation, a footnote may be used to (a) provide a brief explanation or definition; (b) provide additional information; or (c) refer to another page, section, or chapter.

2. *Numbered List of References*
Reports are sometimes documented with a numbered list of references that have been indicated by a number placed in the text. The number may be inserted as follows:

The initial cost of water-saving plumbing devices was about $200 per dwelling.(1)

To clarify the purpose of the number in parentheses, the writer can use an asterisk after the first number in the text, and then provide a footnote explaining that the numbered list of references will be found at the end of the text.

The reference list itself is presented in the order of the numbers in the text. The forms are similar to those for footnotes, except that (a) the publisher's name and address are not enclosed in parentheses, and (b) page numbers are given only for shorter items that appear in longer works (as in no. 4 in the sample list below). Using the list, the writer can cite references as often as needed by merely repeating a number. Here is an example of the list following the text:

References

1. Thomas P. Johnson, Analytical Writing, New York: Harper & Row, 1966.

2. John P. Jones, Modern Business Writing, 3rd ed., Englewood Cliffs, N.J.: Prentice-Hall, 1975.

3. "Managers Go Back to School," Sacramento Bee, 5 March 1979.

4. John D. Roberts, "Management by Committee," in Readings

References (cont.)

<u>for Managers</u>, ed. Thomas P. Bancroft, New York: Harper & Row, 1973, pp. 241-245.

5. Willard Thomas, "Use of Video in Technical Training," <u>Technical Communication</u>, 26, No. 4 (1979), pp. 4-5.

6. Robert Wilson and Michael A. Smith, <u>Communicating with Readers</u>, New York: McGraw-Hill, 1980.

3. *Bibliography*

A general bibliography includes all material on which the writing was based. It differs from footnotes in that it does not document specific facts; rather, it shows all the sources from which the writer drew information. The forms for bibliographic entries are similar to those for footnotes with these exceptions:

a. The author's last name is given first to enable the writer to make an alphabetical list.
b. Both the author's name and the title are followed by periods.
c. Publication data are not enclosed in parentheses.
d. Page numbers are included only for shorter items that appear in longer works.

If articles without authors' names are listed, the title takes its regular place in the alphabetical list. If a second work by an author is listed, a solid line (ten hyphens on the typewriter) is used in place of his or her name.

Here is a brief sample bibliography. As suggested by the Chicago *Manual*, the entries are typed single space, and succeeding lines are indented two spaces. (In MLA style [see p. 113] the entries are typed double space, and succeeding lines are indented five spaces.)

Bibliography

Fear, David E. <u>Technical Communication</u>. Glenview, Ill.: Scott Foresman, 1977.

———. <u>Technical Writing</u>. 2nd ed. New York: Random House, 1978.

Johnson, Thomas P. <u>Analytical Writing</u>. New York: Harper & Row, 1966.

"Managers Go Back to School." <u>Sacramento Bee</u>. 5 March 1979.

Sunset Western Garden Book. 3rd ed. Menlo Park, Calif.: Lane Magazine and Book Co., 1979

Thomas, Willard. "Use of Video in Technical Training." Technical Communication. 26, No. 4 (1979), p. 4.

Thompson, Thomas T. Success Isn't Everything. New York: McGraw-Hill, 1973.

DUAL, DUEL. To keep these two straight, simply remember that dual, which refers to two (as dual controls), has the same ending as *plural*. Duel refers to a conflict between two persons, groups, or ideas.

DUPLICITY, DUPLICATION. The first word of this pair denotes deceit or double dealing. Don't be trapped by their similarity in appearance, as was the legislator who spoke of "the costly *duplicity* of effort by two government agencies." The word he meant, of course, was duplication. In this case, he meant that money was being wasted by two agencies doing the same work.

E

EACH. As a singular pronoun, *each* is often inadvertently, and incorrectly, given a plural verb. This frequently occurs in such constructions as "Each of us are going alone." Remember that each is *always* singular:

1. Each of the girls is to receive an award.
2. Each girl is working up to the limits of her ability.

EACH AND EVERY. A redundant expression—really a cliché—that deserves an ignominious plot in the graveyard of worn-out locutions. *Each and every* does not mean *all,* which is plural. Both *each* and *every* are singular, and thus the construction is totally redundant.

EACH OTHER, ONE ANOTHER. Opinion is divided on the use of *each other* in constructions involving more than two persons or things. When three or more are involved, some writers prefer *one another*. The distinction, however, is really a matter of personal preference, and either form is acceptable regardless of the number of persons or items involved:

The ten visitors, all from different countries, greeted *each other (one another)* with apparent friendliness and natural curiosity.

EFFECT, AFFECT. See **Affect, Effect.**

E.G. An abbreviation of the Latin *exempli gratia,* which means for example. For the correct usage of *e.g.*, see **Abbreviations.**

EITHER. *Either* should be used to indicate one of two items, persons, etc. However, *either* in the sense of *each* or *both* is imprecise and should not be used in technical writing. For example: "The land on *either* side of the mountains is flat and barren." In that sentence, *each side* or *both sides* would convey the idea more precisely.

EITHER . . . OR. See **Correlative Conjunctions.**

ELECT (as a suffix). See **Prefixes and Suffixes.**

ELLIPSIS. A punctuation of three periods separated by spaces is called an ellipsis. When words are omitted from a quotation, the ellipsis is used to indicate the omission. If the omission occurs at the end of a sentence, the final period is added. A statement from the preceding discussion of **Either** might be quoted as follows: ". . . *either* in the sense of *each* or *both* is imprecise. . . ."

ELLIPTICAL CONSTRUCTION. A construction is called *elliptical* when it lacks one or more words that it would contain in formal English. Elliptical constructions are customary in conversation but can sometimes lead to misunderstanding when used in writing. For instance, you might say:

1. The instructor said *that* all themes are to be typewritten.
2. The instructor said all themes are to be typewritten.

Although the omission of *that* in the second sentence would not affect the meaning, consider this statement: "He ordered a copy of the report sent to the director's office." Did he order (1) a copy of the report that *had been sent* to the director's office, or (2) that a copy *be sent* to the director's office?

If you agree that technical writing and other exposition should be clear and unambiguous, you will probably be wary of elliptical construction.

EMINENT, IMMINENT. *Eminent* means prominent, or outstanding in performance or character. *Imminent* denotes that something is impending or about to happen.

EMPLOY. Employ is a rather inflated synonym for *use* that is probably better forgotten. "Employ an end wrench to remove the bolts" sounds a bit pompous, wouldn't you agree?

ENGAGED IN. The phrase *engaged in,* frequently used to inflate sentences, is usually meaningless and redundant. For example:

Inflated: They are engaged in a study of light waves.
Concise: They are studying light waves.

Note that the second version eliminates two prepositional phrases and is actually more precise.

ENORMITY, ENORMOUS. *Enormity* does not denote mere size or extent; it is a noun meaning gross evil or wickedness—outrageousness. For example: "The jurors were shocked by the enormity of the defendant's crime." *Enormous,* of course, is an adjective meaning immense.

ENTOMOLOGY, ETYMOLOGY. Take care with this pair and note how each is spelled. Each denotes a *study of,* but the studies are quite unrelated. An *entomologist* studies insects; an *etymologist* studies words, particularly the origin of words.

ENUMERATION. Enumeration within sentences or in sentence style is used to clarify paragraphs that, because of the complexity of the subject, might be misunderstood. An enumeration might also be used within a sentence to lend equal emphasis to all elements of the predicate. When the enumerations are on separate lines, a period or semicolon is used at the end of each element, unless each is very brief and not a complete sentence.

Enumeration, sometimes called listing, is common in technical writing. The following examples show the correct uses of enumeration. The construction is particularly useful when complex information is being presented; the numbers or letters emphasize items of equal importance, thus helping ensure that no important point is missed by the reader.

A. **Within sentences:**

Processes now being studied include (1) biological treatment for removal of nitrogen, a nutrient that stimulates the growth of algae and other aquatic plants; (2) separation techniques, such as adsorption, ion exchange, and chemical precipitation; and (3) membrane techniques, such as electrodialysis and reverse osmosis.

Note: When the introductory statement is a complete sentence, it is followed by a colon:

The repair kit contains three items: (1) spare bolts, (2) extra wiring, and (3) complete instructions.

B. **Within a sentence on separate lines:**

For convenience, these waste-treatment systems will be discussed as:

1. large facilities—those treating 5 million gallons or more per day;
2. medium-size facilities—those treating 1 to 5 million gallons per day;
3. small facilities—those treating fewer than 1 million gallons per day.

Note: At the writer's option, "bullets" may be used instead of numbers:

The costs of reclaimed water vary widely. Present costs range from:

- $2 to $5 per acre-foot in areas near a treatment plant;
- $20 to $40 per acre-foot when extensive treatment, storage, and transportation are required;
- as high as $100 per acre-foot when extensive treatment, such as desalting, is required.

C. **Sentence style:**

The nozzles were modified in three ways: (1) V-44 rubber was used in the spherical gaps. (2) The tungsten throat inserts were cut back 0.030 in. at the forward ends. (3) The flow straighteners were modified to increase the velocity of the gas flow forward of the throat.

D. **Full sentences on separate lines:**

Three reasons for consideration of the higher limit were listed:

1. Waste water with a high dissolved-solids content might be treated and blended with high-quality water to produce usable water supplies.
2. Waste water containing excessive dissolved solids might be demineralized.
3. Mineralized waste water is suitable for certain uses, such as industrial cooling and firefighting.

ESTIMATED AT ABOUT. Avoid this redundancy. An estimate is not expected to be precise and therefore implies "about" or "approximately":

Redundant: The cost of the bridge is estimated *at about* $2 million.

Revised: The cost of the bridge is an estimated $2 million. (OR) The bridge will cost about $2 million.

ETC. Writers should not use *etc.* (Latin *et cetera*) to hide the fact that they haven't done their homework or because they have run out of things to say. When *etc.* appears at the end of a series of unrelated items, readers are left up in the air. For instance: "The required printing supplies include four trays, two pairs of tongs, a timer, a thermometer, developer, etc." Reading that, the beginning photo buff would be lost.

Use *etc.* only when the implication is obvious. For example: "The benchtop cabinet is ideal for storing nuts, bolts, washers, brads, etc." As a rule of thumb, don't use *etc.* unless you have named at least three related items.

EUPHEMISM. A euphemism is the substitution of a word, a phrase, or even a sentence for one that may seem harsh, unpleasant, or offensive. A euphemism is used to cushion the blow of a more direct word or statement. Some common examples include "detain for questioning" instead of "arrest"; "demise" instead of "death"; "previously owned" instead of "used" (car). Here are a few more examples:

Euphemism: It may be necessary to impose more stringent fuel-conservation marketing policies.
Direct: We may have to increase the price of gasoline.

Euphemism: Temporary sanitary kiosks will be installed at all campgrounds.
Direct: We will install portable toilets at all campgrounds.

Euphemism: We are revising our priorities in order to complete your work by June 1, only a few days beyond the original target date.
Direct: I know you wanted your work finished by May 10. Unfortunately, we can't complete it until June 1.

As you can see in each example, the writer, trying to be tactful, at least succeeded in evading the issue. Except in certain **correspondence,** where tact is sometimes advisable, technical writers should avoid euphemisms and say exactly what they mean.

EVERYBODY, EVERYONE. Both *everybody* and *everyone* denote a singular meaning and should not be used with a plural adjective (possessive pronoun). When the *one* in *everyone* is stressed, it is written as two words.

1. Everybody who wanted a ticket raised a (not *their*) hand.

2. Everyone in the room left with a smile (not *with a smile on their face*).
3. *Every one* of the girls received an award.

EXAMPLES, USE OF. See **Definitions.**

EXCLAMATION MARK. An exclamation mark is used after an interjection, or exclamatory word or clause:

1. Aha! At last!
2. Now I have the answer!

Not all interjections merit an exclamation mark; some may be followed by a comma or period.

1. Oh, come on now.
2. Well, I guess so.

Note: The use of interjections is probably more pertinent to fiction than to technical writing.

EXPLETIVE. An expletive, or anticipatory subject, is an introductory word—*there* or *it*—used to open such sentences as "There are three plans that must be studied." Here the expletive stands as the grammatical subject, whereas the actual subject is in the predicate. The real meaning, of course, is, "Three plans must be studied." Note in the following examples how the explctive results in a weak, indirect statement:

Weak: It was estimated by the manager that the job would last six weeks.
Direct: The manager estimated that the job would last six weeks.

Weak: It is mandatory that the rules be followed.
Direct: The rules must be followed.

Weak: There is one problem that remains to be solved.
Direct: One problem remains to be solved.

On the other hand, an expletive is quite useful and natural in such statements as, "There is no other answer," "It may rain tomorrow." It is also useful in such introductory statements as:

1. There are two principal reasons for our decision. One is . . .
2. There are several ways to check spark plugs. One method . . .

Occasionally, an expletive may be preferable to an awkward attempt to avoid it. In his biography of Sinclair Lewis, Mark Schorer quotes what he calls "one of the great English sentences" (attributed to Calvin Coolidge): "No necessity exists for becoming excited."[10]

EXPOSITION, EXPOSITORY WRITING. Exposition is writing that primarily informs or instructs. Technical writing is classed as exposition. See *Technical Writing*, part 1, p. 4.

EXTRA (as a prefix). See **Prefixes and Suffixes.**

F

FACTOR. Factor is an overused, vague word that (1) often means *part,* (2) often is used in place of a more specific word, and (3) often is meaningless. For instance:

1. **Meaning part:**

   Original: An important *factor* in our planning is the preservation of wild rivers, lakes, and wilderness areas. An equally important *factor* is the protection of wildlife.
   Revised: An important *part* of our planning is the preservation of wild rivers, lakes, and wilderness areas. Equally important is the protection of wildlife.

2. **Vague meaning (in place of a more specific word):**

   Original: Another negative *factor* is the high transportation cost.
   Revised: Another *disadvantage* is the high transportation cost.

3. **Meaningless:**

   Original: An important *factor* to be taken into consideration is that the productivity of a water well may eventually decline.
   Revised: The productivity of a water well may eventually decline.

FARTHER, FURTHER. Most authorities agree that *farther* should be confined to expressions of physical distance, whereas *further* should be used to denote abstract relationships of degree or quality. For example:

1. How much farther is it to John's house?
2. Cleveland is 100 miles farther south.

Note: Further is sometimes used in the sense of distance: "It is 10 miles further to Santa Fe." However, *farther* is considered substandard in the sense of degree or quality:

[10]Mark Schorer, *Sinclair Lewis: An American Life* (New York: McGraw-Hill, 1961), p. 53.

3. Nothing could be *further* (not *farther*) from the truth.
4. The committee will meet tomorrow to consider the question *further* (not *farther*).
5. If you use too many credit cards, you'll get *further* (not *farther*) in debt.

FEET, FOOT. Although *feet* is simply the plural of *foot,* some writers run into trouble with such expressions as *a rope 3 feet long.* The correct usages of *feet* and *foot* are as follows:

1. a *5-foot* plank
2. a fence 6 *feet* (not *foot*) high
3. a *6-foot-high* fence
4. The fence is 6 *feet* (not *foot*) high.
5. I need 120 *feet* (not *foot*) of wire.
6. The wire is sold in *50-foot* (not *feet*) rolls.

FEWER, LESS. *Fewer* is used with items that can be counted or viewed individually, as:

1. *Fewer* than twenty refrigerators were sold last week.
2. *Fewer* than one hundred people attended the lecture.

Less is used to refer to quantities considered in bulk or overall size, as:

1. This year we have had *less* snow than last year.
2. The new model has considerably *less* leg room than last year's.

FIGURES. See **Graphic Aids.**

FLAMMABLE, NONFLAMMABLE are the preferred adjectives to denote "capable (or incapable) of burning." The older form *inflammable* is now considered ambiguous and misleading. The noun form (for flammable) is *flammability.*

FOLD (as a suffix). See **Prefixes and Suffixes.**

FOOTNOTES. See **Documentation.**

FORCED, FORCEFUL, FORCIBLE. Each of these three adjectives differs in meaning, and writers should use them correctly:

1. *Forced* indicates an action or condition imposed by external influence, as a *forced* landing or *forced* labor.
2. *Forceful* implies a potential for force or strength, or displaying those qualities, as a *forceful* speaker or a *forceful* personality.
3. *Forcible* means accomplished by force, as a *forcible* entry (where, for example, a door is broken open or a window broken).

FOREIGN WORDS. Foreign words and phrases that have not been anglicized are commonly underlined on the typewriter and italicized when set in type for printing. At times, especially in formal writing, certain foreign words are punctuated with diacritical marks to indicate their pronunciation (see **Diacritical Marks**). This is particularly true of French words, such as *crêpe, hors d'oeuvre, tête-à-tête, vis-à-vis.*

On the other hand, a great many foreign words have now become so common in English that they are no longer italicized (or underlined) except, perhaps, in scholarly journals. A rule of thumb: do not underline such words unless they are likely to be misunderstood. This, of course, is a matter of the writer's judgment. Most readers would have little trouble with such words as cafe, chic, cliché, éclair, lingerie, quasi, rendezvous, and salon. On the other hand, *nouveau riche, sine qua non, nom de plume,* and *coup d'etat* would be more difficult for many readers.

The University of Chicago *Manual of Style* suggests that whether a foreign word is to be italicized is up to the editor or writer. When in doubt, the *Manual of Style* suggests, do not underline.

See also **Underlining; Diacritical Marks.**

FORTH, FOURTH. Don't be trapped by this pair, as was the student who wrote of the *fourthcoming election.* Simply remember that *fourth* follows third, and you'll remember the correct use of the adverb *forth* (meaning forward in time, place, or order). One of the anomalies of the English language, by the way, is that *four* has a *u* but *forty* has none.

FORTUITOUS, FORTUNATE. A *fortuitous* event is one that occurs entirely by chance; that is, it is completely unplanned and unforeseen. *Fortunate* simply means lucky (so, of course, a *fortunate* event might also be *fortuitous*).

For example, if you were unemployed and actively seeking a job, your finding one would be a *fortunate* event. Yet it would not be *fortuitous,* because it resulted from your planning and active job search. On the other hand, if your car suddenly stopped running while you were in front of a service station, it would be a *fortuitous* event. It would also be *fortunate* for you that you were passing the service station.

FRACTIONS. In text, fractions should be written out; i.e., figures would not be substituted for the words:

Wrong: Almost ¾ of the students were absent.
Right: Almost three-fourths of the students were absent.

See also **Numbers.**

FRAGMENTARY SENTENCE. See **Sentence Types and Construction.**

FROM WHENCE. Another redundant expression to avoid. *Whence* means from where or from what place; therefore, *whence* says it all. (He returned *whence* he came.)

FULL, FULSOME. Be careful not to use *fulsome* when you mean abundant; *fulsome* means insincere or offensive to good taste. *Fulsome* pertains chiefly to flattery or overdemonstrative praise or adulation.

G

GENDER. See **Sexist Language.**

GERUND. A gerund is a verbal **noun,** more accurately, the present **participle** of a verb used as a noun:

1. *Swimming* is good exercise.
2. *Writing* is hard work.

See also **Participle.**

GET, GOT. *Get* and *got* are used colloquially and idiomatically in the sense of arise, arrive, become, buy, call, catch, finish, go, have, hurry, leave, manage, meet, must, progress, receive, and understand, to name only seventeen. For example:

1. I got up (arose) at 6:30 this morning.
2. I got there (arrived) just in time.
3. I got (became) nervous waiting for the test to begin.

4. I got (bought) a new car last night.
5. I'll get hold of (call) you tomorrow.
6. I got (caught) my limit of trout in an hour.
7. I'll get it done (finish it) tomorrow.
8. I must get (go) to work.
9. I've got (I have) ten dollars. How much have you got?
10. You had better get with it (hurry).
11. I must get going (leave).
12. I'll get by (manage) somehow.
13. We're all getting together (meeting) at John's house.
14. I've got to (I must) leave now.
15. We don't seem to be getting anywhere (progressing).
16. I got (received) $200 for that old picture.
17. I don't get it (understand).

In conversation, those and other uses of *get* and *got* seem perfectly natural. In technical writing, however, writers should say what they really mean. Too many *gets* and *gots* will lend a **colloquial** flavor—and sometimes the flavor of **slang**—to otherwise precise writing.

GOBBLEDYGOOK. The *American Heritage Dictionary* defines this ungraceful word as "bureaucratic jargon (from GOBBLE [to sound like a turkey] influenced by GOOK)." If that doesn't tell you enough, I will add that in gobbledygook, **verbs** seldom act (they remain passive), two words usually take the place of one, and the **diction** is **pompous,** frequently outmoded, and often obscure. Here are a few examples:

Gobbledygook: An energy-conscious awareness is to be maintained by all personnel.
Improved: All employees should try to save energy.

Gobbledygook: Once fair and full consideration has been given to competing testimonies, a decision will be announced by the council.
Improved: The council will hear testimony from both sides and then announce its decision.

Gobbledygook: Attitudinal responses based on considered concepts of environmental values were sought by the researchers.
Improved: The research team asked people how they felt about the environment.

Gobbledygook: Evaluation of the petition will be conducted in due course.
Improved: We will evaluate the petition as soon as possible.

In an effort to wipe out gobbledygook, one enterprising manager wrote a brief memo asking all employees to use plain language. He concluded it with, "After all, a spade is a spade is a spade. Let's stop calling it a trenching device."

See also *Say What You Mean—Clearly* (part 1, p. 13); **Inflated Expressions; Pompous Language.**

GOOD. As is true of **get** and **got,** *good* is used conversationally to denote a large number of meanings. Here are twenty-eight; undoubtedly this list could be longer:

a good (ample) income
a good (at least) 6 miles away
a good (beneficial) rainfall
a good (plentiful or bountiful) harvest
a good (brisk) wind
a good (comfortable) chair
a good (complete or thorough) physical examination
a good (competent) laboratory technician
a good (delicious) steak
good (discriminating) taste (in clothing)
good (efficient or effective) organization
a good (enjoyable) trip
a good (favorable) omen
good (genuine, not counterfeit) money
a good (informative) textbook
a good (loyal) American
a good (nutritious) breakfast
good (pleasant) weather
good (potable) water
a good (sound) investment
good (healthful) exercise
a good (long-lasting) paint job

a good (thorough) reprimand

a good (logical) explanation

a good (valid) reason

good (well-made) furniture

As with other colloquial expressions, the meaning of *good* in conversation is usually obvious. On paper, however, technical writers should use words that convey precise meanings.

GOOD, WELL. *Good* is often confused with *well* and incorrectly used as an **adverb.** In conversation, you often hear such statements as "The experiment went real good yesterday," or "The furnace is working good today." In those examples, *good* incorrectly modifies a verb, but in informal conversation some people don't mind. In technical writing, however, *good* should be used as an adjective; *well* should be used when an adverb is called for:

Good: She wrote a *good* theme (adjective—modifies *theme*).

Well: She writes *well* (adverb—modifies *writes*).

Good: He presented a *good* speech (adjective—modifies *speech*).

Well: Her speech was presented *well* (adverb—modifies *presented*).

Good and *well* also have different meanings in the sense of well-being. That is, you can feel *well* (not ill), or you can feel *good* (about passing an examination, for example).

GRAPHIC AIDS. In almost all technical writing, and particularly in reports, writers will find use for graphic aids—tables, figures, and photographs. Using a table, for example, a writer can save a great many words and eliminate needless repetition. Figures, including various types of charts and graphs, help readers visualize what they are reading. Photographs are also valuable aids to understanding; they let readers see what they are reading.

Tables

Writers should be aware of the advantages of tabulating information, particularly in discussions containing a great many numbers. Strings of numbers are difficult for readers to follow and sometimes can become almost incomprehensible. Here are several examples.

How Tables Save Words and Increase Comprehension. Consider the following paragraph from a discussion of water quality:

The proportion, or concentration, of dissolved minerals (salts) in water is a determining factor in evaluating its quality. The average concentration of salts in fresh water is so small that it is expressed as milligrams per liter (mg/l). For example, a concentration of 100 mg/l means that 1 liter of water contains 100 mg of salts. If you added 1 mg of table salt to 1 liter of pure water, you would have a concentration of 1 mg/l.

The dissolved mineral content of rivers is usually lower than 500 mg/l, although some rivers may contain as much as 2,000 mg/l. Distilled water for automobile batteries is salt-free, i.e., 0 mg/l. Examples of different mean salt concentrations in various sources of water include: distilled water, 0 mg/l, rain, 10 mg/l; Lake Tahoe, 70 mg/l; Sacramento River, 150 mg/l; Lake Michigan, 170 mg/l; Pacific Ocean, 35,000 mg/l; brine well, 125,000 mg/l; Dead Sea, 250,000 mg/l; Great Salt Lake, 266,000 mg/l.

Now, let's convert the second paragraph to a summary statement and a table. We can change the third sentence in the paragraph as follows:

Examples of different mean salt concentrations in various sources of water are shown in table 1.

TABLE 1 **Mean Salt Concentrations in Various Sources of Water**

Source	*Mean salt content (mg/l)*
Distilled water	0
Rain	10
Lake Tahoe	70
Sacramento River	150
Lake Michigan	170
Pacific Ocean	35,000
Brine well	125,000
Dead Sea	250,000
Great Salt Lake	266,000

The advantages of the tabular presentation for both writer and reader should already be clear. First, the table headings eliminate the need for the constant repetition of the unit (mg/l). Given once in the table heading, the unit applies to each number shown in the body.

Second, readers can see each source and salt content on a separate line, enabling them to compare and determine the different salt concentrations in the various sources at a glance. By contrast, with the numbers strung out in the rather dense paragraph, the significance of the differences in salt content may escape them altogether. Yet, from the table, they can quickly determine that the average salt content of Great Salt Lake, for example, is almost eight times that of the Pacific Ocean.

In some instances, writers may have to present information that simply could not be put into words without a long, repetitive explanation. Consider, for example, this table comparing the merits of four automobiles:

TABLE 2 **Four Automobiles: Comparison of Driving Ease**

Test	*Automobile*			
	A	*B*	*C*	*D*
Acceleration	Smooth	Slow	Smooth	Smooth
Ease of handling	Superior	Fair	Good	Excellent
Level ride on rough surface	Average	Noticeable sway	Comfortable	Average
Comfort of ride, 55 mph (highway)	Medium	Medium	Excellent	Good
Noise level, 55 mph (highway)	High	Average	Low	Low

Now visualize trying to write a paragraph, or several paragraphs, comparing the results of the five tests applied to four automobiles. Where would you begin? How many words would it take? And even if readers managed to struggle through it, could they possibly see the comparisons shown so clearly in the table?

Finally, here is a more complicated table with three columns of numbers. As you can see, this information simply could not be presented verbally:

Table 3. Projected Urban Per Capita Water Use by Hydrologic Study Area, 1967-2020 (gallons per day)

Study Area	1967	1990	2020
North Coastal[1]	160	140	130
San Francisco Bay	170	200	220
Central Coastal	200	210	210
South Coastal	180	190	200
Sacramento Basin[1]	350	350	350
Delta-Central Sierra[2]	320	280	260
San Joaquin Basin	370	390	420
Tulare Basin	370	350	350
North Lahontan	Per capita values not available		
South Lahontan	280	320	320
Colorado Desert	380	400	400

[1] Water demands for pulp and paper industry not included.

[2] Values are for valley floor only. Recreational and second-home use in Sierra foothills not included.

Another advantage of a numbered table is the ease with which it may be referred to at any time. For example, assume that you have presented table 1 on page 5 (of a report) and that you want to refer to it again on page 11. All you need do is to refer to table 1, eliminating the need for an awkward reference, such as: "As shown in the table on page 5," or "As shown in the preceding compilation." In addition, when you identify the table by number, you obviate the need to refer to it by title, a great convenience for both you and your readers. Still another advantage of numbered and titled tables is that you can list them in the table of contents. It is always advisable, particularly in long reports with many tables and figures, to give each a number and title.

Construction and Use of Tables in Text. Here are a few standard rules for use of tables.

1. Number the tables in order of presentation: Table 1, Table 2, Table 3, etc.
2. Give each a title, but in the text, refer to them by number only: "Table

1 presents"; "As shown in table 1"; "The data in table 1 show"; etc. Since the title is above the table, there is no need to repeat it in the text.

3. Present each table as close as possible to its first reference. Often you will be able to present a small table on the same page as the reference.
4. Try not to present a table until you have referred to it. If you present the table on page 5 but don't mention it until page 7, readers will wonder about its relevance.
5. Each column should have a heading; determine the width of the columns by the width of the data in each. To save space, you may use common abbreviations and symbols in the column headings; unusual abbreviations, unless clarified in the text, should be explained in a footnote to the table.
6. When applicable, include a unit in the table heading to avoid repeating the unit in the body. For example: "Cost ($)"; "Length (in.)"; "Weight (tons)." To avoid unwieldy numbers in the body, express quantities in the highest common multiple, such as "thousands of acres" or "acres × 1,000." However, avoid scientific notation, such as "acres × 10^3." If one unit applies to all the data, it may be included with the title (as in sample table 3, p. 134). Here is a typical table heading:

G

TABLE 1 Steam-Electric Generating Plants Owned by California Utilities

Name	*Location (County)*	*Capacity (MW)*	*Type of Cooling*	*Fuel*
Rancho Seco	Sacramento	913	Cooling towers	Nuclear

7. To keep figures on a single line, don't use fractions in the body; use decimals instead. Line up decimals on the decimal point. When there are decimal and whole numbers in the table, add a zero or zeros to each whole number to equalize the columns; precede each decimal number (less than 1) with a zero. For example:

Mineral	*Mg/l*
Copper	1.00
Iron	0.30
Sulfur	0.05
Manganese	0.77
Potassium	5.00

8. A footnote to a table should be presented in the table to distinguish it from ordinary footnotes in the text (as in sample table 3, p. 134).
9. If a table continues on the next page or pages, add "Continued" in parentheses after the number and title.
10. Tables should be ruled horizontally and are often ruled vertically as well (as in the sample table 3, p. 134). A modern trend is to rule vertically only the column heads and to leave "white space" between the columns in the body (as in sample table 2, p. 133).

See also the tables in the sample report (**Reports,** p. 236ff.).

Figures

Figures have a different mission. Rather than substituting for words, figures supplement a writer's words; figures let readers see what they are reading. Although it is often true that one picture is worth a thousand words, figures do not usually completely replace words; rather, they illustrate the writer's words for the reader's benefit.

Figures you should be familiar with include bar charts, circle, or pie, charts, flow diagrams, and graphs. As a writer, you will not have to draw them—that task is usually the province of a skilled delineator—but you should know how to use them.

Bar Charts. Bar charts help readers compare different sizes and quantities being discussed in text. They are especially helpful when you are discussing numbers or sizes that are either close together or far apart. That is, readers may miss the significance of the difference, say, between 0.5 and 1.6 (close) or 12 million and 14 million (2 million difference) until they see the difference emphasized by the length of the bars in the chart.

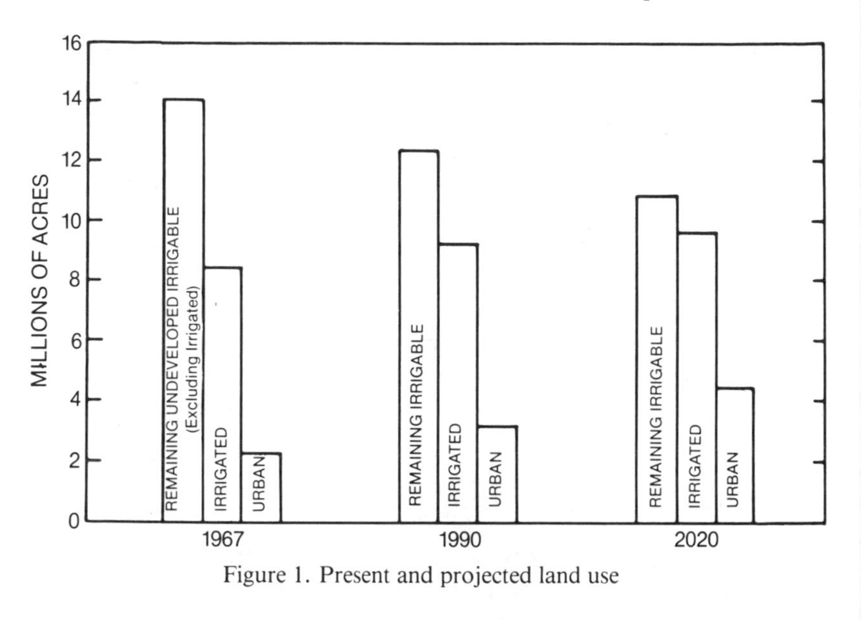

Figure 1. Present and projected land use

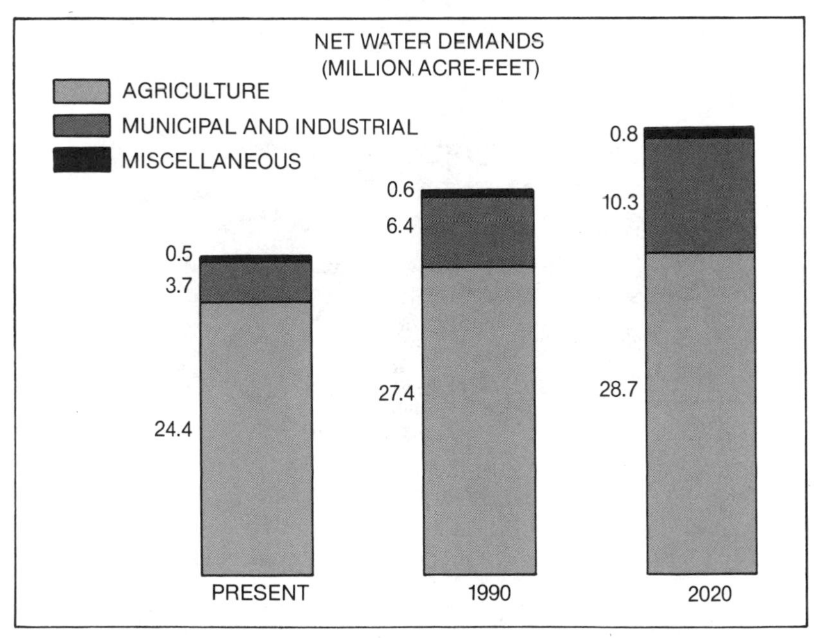

Figure 2. Projected growth of net water demands, 1967–2020

In some cases, the meaning of the bars will be differentiated by lettering (as in sample figure 1).[11] In other charts, bars may be differentiated by colors or shades (screens) of black (as in sample figure 2). In the latter case, the chart will require a legend explaining the different colors or shades of black. The bars can also be differentiated by giving each a distinct pattern, e.g., diagonal lines, crosshatching, or a "Zipatone" pattern. (Zipatone is a press-on material available in many distinct patterns.)

Pie Charts. Pie charts are useful to show proportions—how much was allocated, say, to expenses, purchases, salaries, etc. They are simple to construct, provided there are not too many divisions in the circle. Sample figure 3 is a simple pie chart that emphasizes an important point the writer

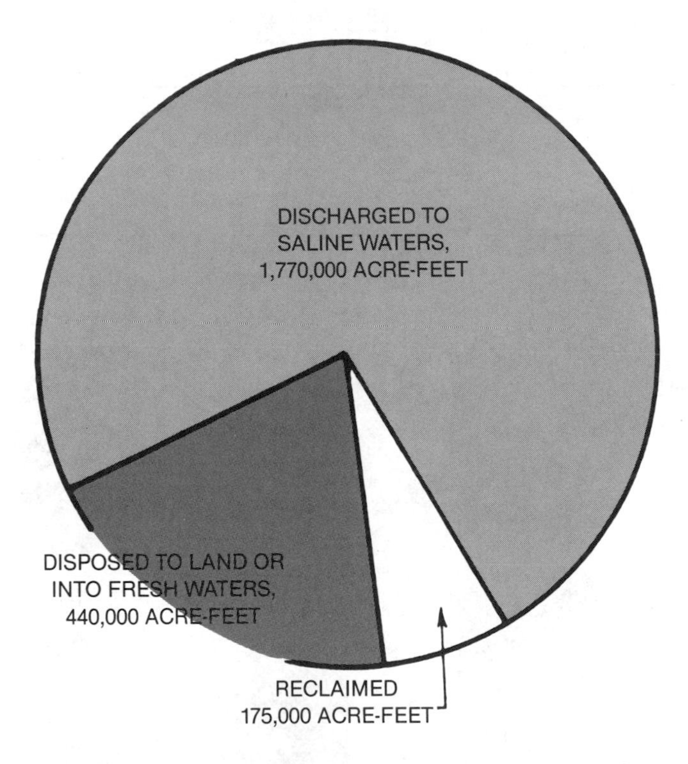

Figure 3. Disposition of urban waste water produced in 1975

[11] I would like to thank the California Department of Water Resources for permitting me to reprint these illustrations (sample figures 1 through 6), that originally appeared in various department reports.

stressed in the text: that almost three-fourths of the waste water treated during 1975 (in California) was discharged to the ocean instead of being reused. Here the chart enabled the writer to highlight the significant difference between the quantity of treated waste water "lost" to the ocean and the quantity reused (reclaimed).

In sample figure 4 the writer used three pie charts to show the estimated proportional growth of water demands over some fifty years. Note that the circles are three different sizes, emphasizing that the total water demand will increase as time passes. The writer also used a legend in this chart to explain the different colors or shades of black.

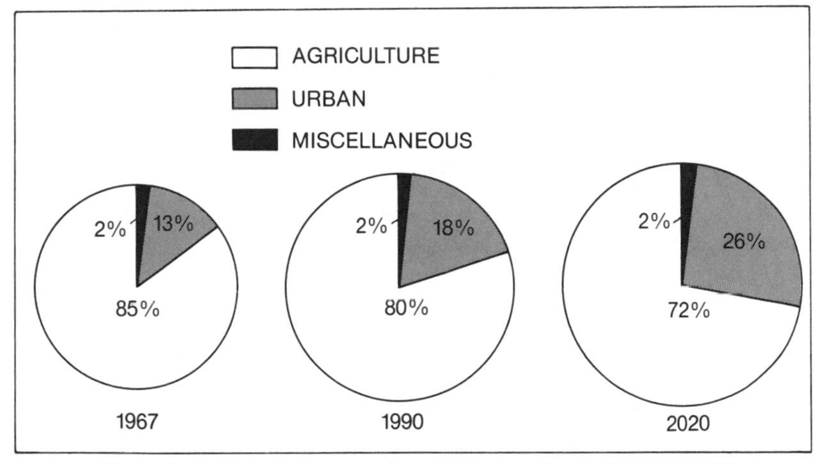

Figure 4. Proportionate growth of net water demands, 1967–2020

Flow Diagrams. Flow diagrams help readers follow explanations of technical processes, which can often be complicated in text. The chart diagrams the process and enables the reader to follow it. Note how the following paragraphs and sample figure 5 complement one another:

Steam Cycle and Flow of Cooling Water

The cooling water system condenses the steam after its useful heat has been expended in the turbine. The cooled water is then returned to the power plant boiler, where it is reheated to produce additional steam.

In power plants that use fossil fuels or nuclear energy, the cooling water system is an integral part of the steam supply system. In the power-generation portion of the plant, the high-pressure superheated steam passes through the

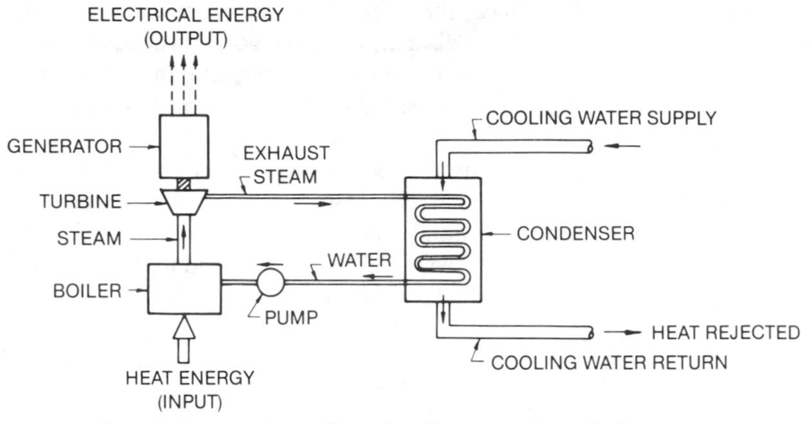

Figure 5. Steam cycle and flow of cooling water, schematic diagram

turbine, imparting energy to the turbine shaft, which turns the generator rotor and produces electricity.

The spent steam then leaves the turbine and passes to the condenser, where it flows over cooling tubes and is changed back to water (condensed). Heat is picked up from the condensing steam and rejected to the atmosphere. The condensate (cooled water) is returned to the boiler (or nuclear reactor) and again becomes high-pressure steam.

Figure 5 traces the basic steam cycle and flow of cooling water.

Graphs. Graphs differ from bar or pie charts in that they show changes in at least two values, one of which depends on the other. For example, you might want to portray how people use more water for lawn and garden sprinkling as the temperature rises in late spring and summer. Or you might want to portray a trend, such as an increase in population with the passage of time. Both of those situations can be plotted on a graph.

To show how changes in one value affect the other value, the graph must have two scales: a horizontal scale (*abscissa*) and a vertical scale (*ordinate*). To indicate the trend correctly, the dependent variable must be plotted along the vertical scale and the independent variable along the horizontal scale. In the two examples mentioned in the preceding paragraph, the increase in watering "depends" on the rise in temperature; the population increase depends on the passage of time.

That information could be presented in a table, of course; however, unless readers pay close heed to the differences in the tabulated data, they may not see the trend, or trends, you want to emphasize. The latter, however, can usually be seen clearly in the curve of a graph.

For example, sample figure 6 shows graphically the increase in irrigated farm acreage in California since 1930 and how it is expected to increase between 1980 and 2020. The curve shows an average increase of 80,000 acres per year from 1930 to 1940. Then the curve rises sharply to show the

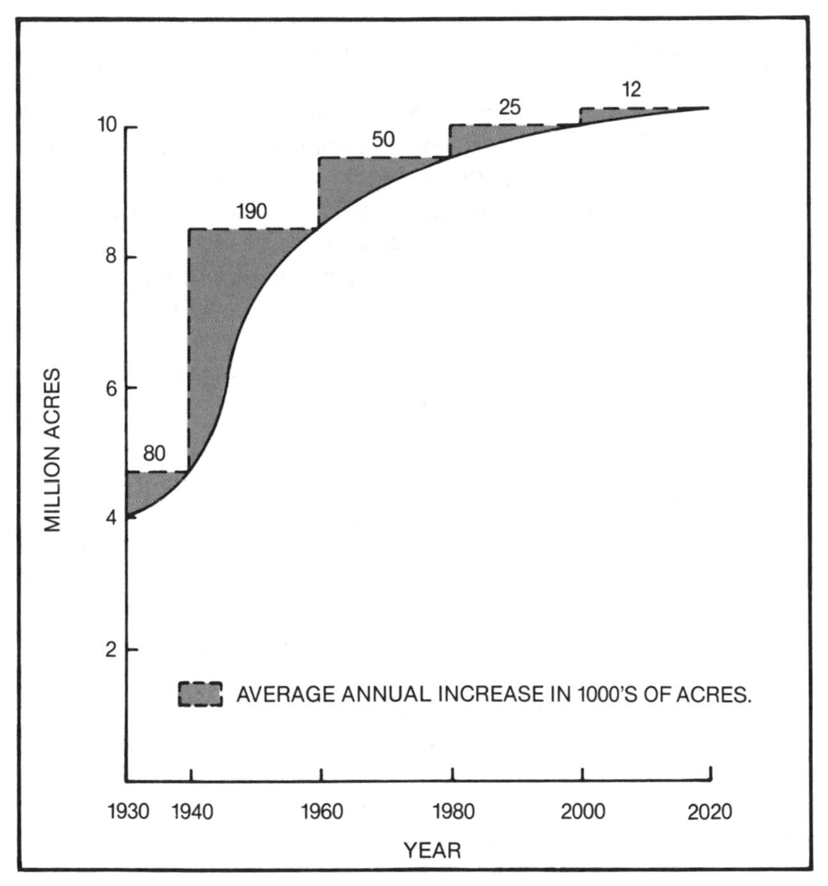

Figure 6. Historical and projected growth of net irrigated acreage, 1930–2020

average increase of 190,000 acres per year between 1940 and 1960, more than twice the rate of annual increase during the preceding ten years.

Between 1960 and 1980 the curve levels off again, showing a modest gain of only 50,000 acres per year. Finally, the curve becomes even more level, showing a projected (estimated) annual increase of 25,000 acres per year between 1980 and 2000 and just 12,000 per year between 2000 and 2020. Thus the graph highlights the large annual increases between 1940 and 1960, a period of tremendous growth in irrigated farm land (in California).

To prepare a graph:

1. Make certain the gradations in both scales are equal. The curve will portray the data accurately only if the gradations are identical.
2. If you plot the points on printed graph paper, make the curve bold (heavy) enough to show clearly. Avoid lettering on the grid. Type or letter any necessary information on gummed labels; cut the labels neatly and apply them over the grid (as in sample figure 7).
3. If it is impractical to start with zero, make certain that readers know the zero has been "suppressed." For example, if all values were between 500 and 1,000, there would be no point in starting with zero. You can indicate the suppressed zero with a break in the vertical scale (as in sample figure 7).
4. If you are showing two or more curves, either label them or use a legend to identify each clearly (as in sample figure 7).

Use of Figures in Reports.

1. Number figures in order of presentation: Figure 1, Figure 2, etc.
2. Give each a title, but in text simply refer to them by number: "Figure 1 presents," "As shown in figure 1," etc. There is no need to repeat the title in the text or to tell readers that "Figure 1 is a graph." Instead of "Figure 1 is a graph of friction vs. time," try, "Figure 1 shows how friction increases with time."
3. Present each as close as possible to your first text reference to it. Often you will be able to present a small figure on the same page as the reference . When a report is to be printed back to back, a figure can usually be presented on the page facing its reference.
4. In figure references, avoid such wording as, "As figure 2 *clearly shows,*" or "As the *reader can easily see* in figure 2." Whereas the figure may be quite clear to the writer, it may not be at all clear to readers, and you will only irritate them with such introductions.

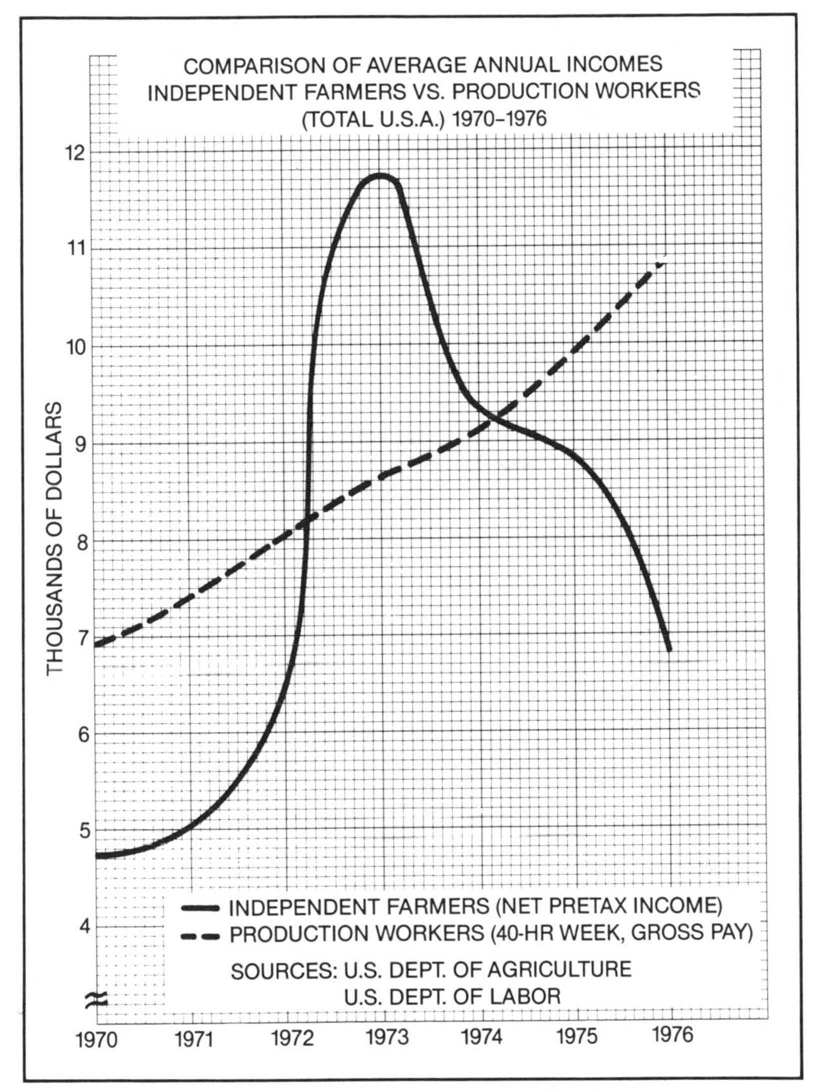

Figure 7. Graph with components identified by legend

Photographs

Whereas photographs are often used to dress up brochures and pamphlets, their principal value in technical reports is that they enable you to show things as they actually exist. For instance, if you were preparing a report on the damage caused by a flood, what better description or evidence could you provide than a sequence of photographs showing the damage: perhaps a washed-out bridge, mud-covered automobiles, streets covered with debris, fallen trees, buildings with silt-covered floors, etc.

When using photographs to portray your words graphically, you should use them to maximum advantage. That is, the photo should literally center on the subject you are discussing. Assume that you wanted to show the elaborate farm sprinkling system shown in the two photographs here. In the first photo, the subject is off to the left and appears to be almost incidental to the vast expanse of sky and field. In the second photo, however, the sprinkler is in the foreground—the photograph now centers on the sprinkling system, the subject described in the *caption* (the words below the photo).

Both photographs are identical, however; that is, they were prepared from the same negative. In the first, the writer merely printed the photo as

PHOTOGRAPH, NOT CROPPED

MECHANICALLY MOVED SPRINKLER SYSTEM

PROPERLY CROPPED PHOTOGRAPH

MECHANICALLY MOVED SPRINKLER SYSTEM

it was "shot," without regard for the importance of the subject. In the second, however, the writer prepared the photo to bring the subject into sharp focus. To do this, he or she simply blocked out the extraneous areas in the foreground, on the right, and at the top—a process called *cropping*. In so doing, the writer brought the sprinkler system into the foreground by cropping out the extraneous areas and leaving just enough sky and farm land to lend authenticity to the photograph.

Cropping, a simple procedure, is often necessary to emphasize the subject of a photograph. To crop a photo, you can prepare a small paper or cardboard frame, moving it about on the photo until you find the exposure that portrays the subject to best advantage.

Graphic Aids at Work

Look at the four pages reproduced from a report by the California Department of Water Resources—Bulletin 189, *Waste Water Reclamation: State of the Art* (1973). In the brief section shown, which explains the advanced waste water treatment process, the writer used a schematic diagram, three figures, and two photographs to help readers understand the discussion. Note that each figure is as close as possible to its reference: on the same page, a facing page, or the page immediately following.

Advanced Treatment

Advanced treatment (Figure 15) may precede, accompany, follow, or replace primary and secondary treatment. When it follows secondary treatment, it is called tertiary. Advanced treatment removes or reduces specific constituents such as nutrients, suspended and colloidal organics, refractory or nonbiodegradable organics, dissolved inorganic minerals, and pathogenic organisms. Unlike the biological processes used for secondary treatment, most advanced treatment methods have been adapted from processes used for treatment of water supplies and industrial wastes.

Before most advanced waste treatment processes can be effective, most of the suspended and colloidal organic matter must be removed. This is most often accomplished by (1) chemical precipitation, using alum or lime as coagulants, (2) graded mixed-media, sand, or diatomaceous-earth filters, or (3) microscreens.

Removal of Nutrients. Special processes may be used to remove nitrogen and phosphorus, the primary nutrients that stimulate the growth of algae and other undesirable organisms. This is particularly important when reclaimed waste water is to be used for recreational lakes. Phosphorus can be removed by chemical precipitation, with lime as the coagulant.

A portion of the nitrogen (that existing as ammonia) can be removed by air stripping in a tower. In this process, lime is added to the water to convert the nitrogen to ammonia in the form of a dissolved gas. The water is then splashed over a series of slats; as the water strikes a slat, droplets are formed, and the ammonia gas escapes from the droplets. The escaping gas is then removed by air circulation. At the South Lake Tahoe reclamation plant, air enters the stripping tower through side louvers and travels horizontally to the center of the tower, where it is pulled upward by a fan.

Ammonia nitrogen can also be removed by columns of the selective zeolite ion exchanger, clinoptilolite. The ion exchanger is periodically regenerated with caustic and salt.

Chemical precipitation and air stripping, combined with mixed media filters, are used at South Lake Tahoe to remove suspended and colloidal organics, phosphorus, and ammonia-nitrogen. At the South Lake Tahoe reclamation plant, nitrogen content is reduced some 30 to 98 percent, depending principally on temperatures. During cold weather, the air stripping process is considerably less efficient. Phosphorus content in the plant influent is reduced by more than 90 percent, and suspended and colloidal organics are almost completely removed.

If time is not important and land is not costly, nutrients can be removed in settling ponds. Over a period of time, algae will use nitrogen and phosphorus to grow additional algal cells. Removal of the algal cells reduces the nutrients in the remaining water. Unfortunately, this process is difficult to control and thus is generally inefficient.

Nitrogen can also be removed through a two-stage operation using nitrification and denitrification. In the first stage, bacteria are used to oxidize ammonia nitrogen to nitrates. In the second, or denitrification, stage, nitrates are converted to nitrogen gas. This stage requires the addition of a nitrogen-free organic substance (such as methanol) on which the bacteria can feed.

Refractory or nonbiodegradable organic substances can best be removed by passing the waste water through beds of granular activated carbon. Activated carbon will also effectively remove tastes and odors. In addition, the process will remove small quantities of nitrates by providing the anaerobic conditions necessary for denitrification, and it will also remove most suspended organic material remaining in the waste water. As an alternative to activated carbon, weak-

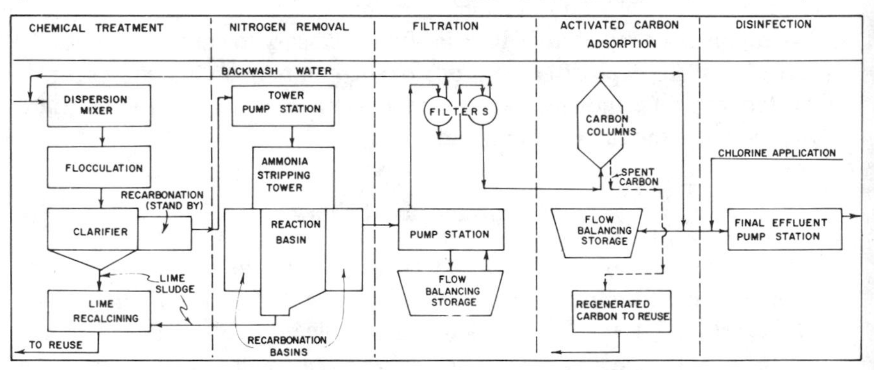

Figure 15. Typical Advanced Treatment Process, Schematic Diagram

Nitrogen Stripping Tower

South Tahoe Public Utility District photo

base synthetic resins have been specifically developed for the removal of soluble refractory organics.

The effluent is then disinfected with chlorine before it is discharged or made available for reuse.

Demineralization. When waste water contains excessive minerals (salts), or when its character may limit reuse, desalting processes may be used to remove dissolved salts.

Electrodialysis combines the use of an electrically charged cell and ion-selective membranes to remove salts from waste water. When a salt dissolves in water, it tends to break down into ions, or small groups of electrically charged atoms, each of which has either a positive charge (called cations) or a negative charge (called anions).

For example, the two parts of common table salt are the chemical elements sodium and chlorine. As salt dissolves in water, the sodium is present as positively charged sodium ions and the chlorine as negatively charged chloride ions.

As mineralized water is passed through an electrodialysis cell, the positively charged ions are drawn through a special membrane to a negative electrode, while the negatively charged ions are attracted through another membrane to a postitive electrode (Figure 16). With the salt ions removed from the influent waste water, desalted water flows out of the cell for reuse.

In actual practice, a number of membranes can be placed between the electrodes, forming a number of dilute (demineralized) and concentrate (waste) compartments. Mineralized water is fed into the electrodialysis cell, and product water and waste concentrates are removed, through manifolds attached to the membrane stacks.

Electrodialysis is used at more than 100 water supply plants around the world, from small industrial plants to a plant in Sarasota County, Florida that treats 1.2 million gallons per day. Here, water with a dissolved solids content of 1,300 parts per million is desalted to a level of 500 parts per million.

Ion exchange is still another demineralization process. An ion exchanger is a porous bed of natural material or synthetic resins that have the ability to exchange ions held in the resin with those in mineralized waste waters that contact the bed (Figure 17).

In the ion-exchange process, both cation (positively charged ions) and anion (negatively charged ions) exchangers are used. The ion-exchange beds are usually placed in series so that the saline waste water passes first through the cation exchanger and then through the anion exchanger.

Carbon Adsorption Columns

South Tahoe Public Utility District photo

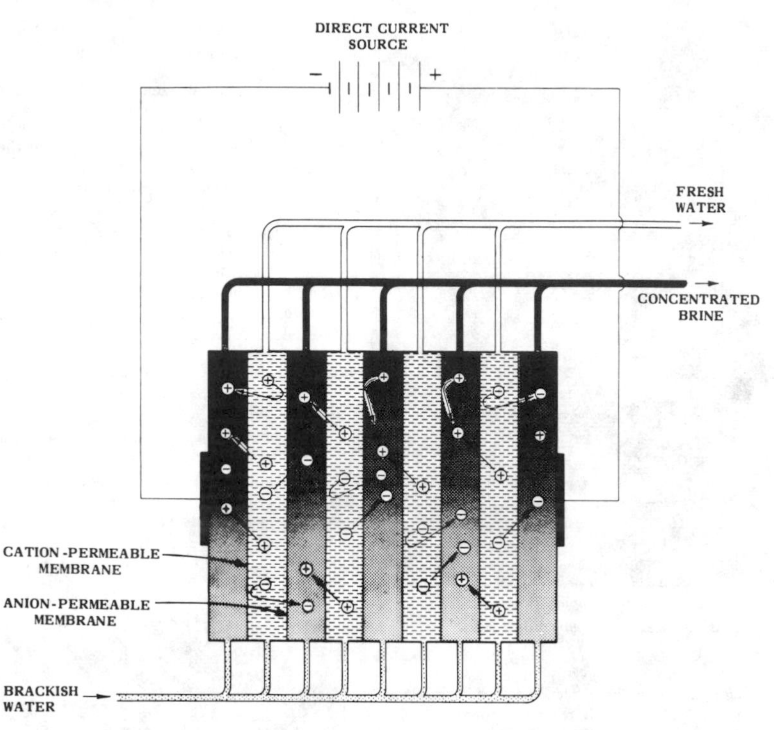

Figure 16. Principle of Electrodialysis

In the cation exchanger, cations, such as sodium, are taken from the salt water, and a hydrogen ion is put into the water. In the anion exchanger, anions, such as chlorine, are taken from the water, and a hydroxide ion is put into the water. Thus, sodium chloride is removed from the waste water, leaving a demineralized effluent. In addition, the hydrogen ion and the hydroxide ion combine to form more water, thus adding to the volume of fresh water produced for reuse or disposal.

As the conversion process continues, the resins become progressively saturated until they finally lose their ability to remove the cations, such as sodium, and the anions, such as chlorine. At this point, the resin beds must be regenerated with acid and caustic to restore their ion-exchange properties. These chemicals used to regenerate the resins also increase the waste requiring disposal.

Ion exchange, reverse osmosis, and electrodialysis are suitable processes for reducing the salinity of brackish waters. Energy requirements for electrodialysis and reverse osmosis are roughly proportional to the amount of minerals removed; i.e., the greater the degree of demineralization, the more energy required. In the ion exchange process, the amount of chemicals used for regeneration is proportional to the amount of salt removed from the waste water.

Reverse osmosis is another demineralization process that may be used. Ordinarily, if fresh water and a salt solution are separated into compartments by a semipermeable membrane, the fresh water will pass through the membrane by osmotic pressure and dilute

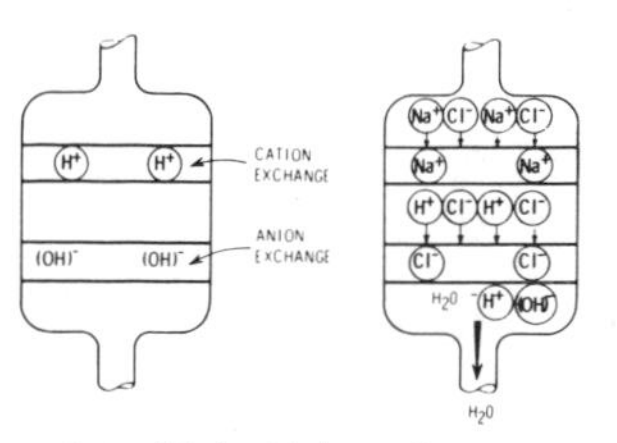

Figure 17. Ion-Exchange Process

the salt solution. However, if pressure is exerted on the salt solution, the osmosis process can be reversed.

When the pressure on the salt solution exceeds the natural osmotic pressure, fresh water from the salt solution will pass through the membrane in the opposite direction to the freshwater side, leaving the salts in a concentrated brine. Thus, instead of drawing pollutants from the water, reverse osmosis extracts water from the pollutants. The process will separate pure water from many waste waters, even those containing bacteria and detergents. Reverse osmosis is a most promising method, particularly for reducing salinity of brackish water. The principle of reverse osmosis is illustrated in Figure 18.

Distillation is still another method that could be used to reclaim waste water. In this process, the mineralized waters are converted to steam and condensed as mineral-free water. However, distillation is mainly used for desalting water with the highest salt concentrations, i.e., sea water, because electrodialysis, reverse osmosis, and ion exchange are less expensive processes for desalting ordinary brackish water.

Variations of the distillation process are used to provide drinking water on ocean liners and in parts of the world where water supplies are limited. Of course, as in any desalting process, the minerals separated from the wastes remain behind and may pose a serious disposal problem.

Costs of Waste Water Treatment

Whether waste water is to be discharged or reclaimed, the costs of treatment vary considerably, depending, among other factors, on climate, size of plant, costs of construction, and quantity and character of the waste water. The estimated annual costs for a number of conventional treatment processes presented in Table 1 are based on typical costs at three different facilities capable of treating 1, 10, and 100 million gallons of waste water per day.

Table 1. Estimated 1972 Costs for Waste Water Treatment (per 1,000 gallons)[1]

Process	Size of facility		
	1 MGD[2]	10 MGD	100 MGD
Primary	$0.15–0.17	$0.07–0.09	$0.04–0.06
Primary with flocculation	0.18–0.20	0.10–0.12	0.07–0.09
Activated sludge[3]	0.24–0.27	0.14–0.16	0.07–0.10
Trickling filter[3]	0.21–0.23	0.10–0.13	0.06–0.08
Carbon adsorption[4]	0.22–0.24	0.10–0.12	0.04–0.06
Chemical coagulation and sedimentation[4]	0.05–0.06	0.04–0.06	0.04–0.05
Ammonia stripping[4]	0.04–0.05	0.02–0.03	0.01–0.02
Mixed media filtration[4]	0.10–0.11	0.04–0.05	0.02–0.03

[1] Includes capital amortization and the costs of normal operations and maintenance.
[2] 1 million gallons per day.
[3] Includes costs of primary treatment.
[4] Not including costs of any previous treatment.

Figure 18. Principle of Reverse Osmosis, Tubular Type (Courtesy of U.S. Office of Saline Water)

GREAT (as a prefix). See **Prefixes and Suffixes.**

H

HALF (as a prefix). See **Prefixes and Suffixes.**

HEADINGS (in reports). See **Reports.**

HEALTHFUL, HEALTHY. Strictly speaking, *healthy* means possessing good health or in good health and almost always refers to a person's physical condition. *Healthful* means conducive to health, or salutary, as a healthful atmosphere or environment.

HOMONYMS. Words that are pronounced alike but that differ in spelling and meaning are called *homonyms.* Unfortunately, they are often confused by writers and used incorrectly. Although the use of the wrong word may not always destroy a writer's meaning, it can seriously impair the impression he or she makes on readers. Here is a list of homonyms that are frequently misused. Each pair or trio listed is discussed in the Handbook.

accept, except
affect, effect
all ready, already
all together, altogether
allude, elude
altar, alter
born, borne
brake, break
capital, capitol
cite, sight, site
complement, compliment
council, counsel, consul
discreet, discrete
forth, fourth
its, it's
lead, led
mantel, mantle
palate, pallet, palette
pore, pour
principal, principle
role, roll
stationary, stationery
their, there, they're
to, too, two
who's, whose
your, you're

Although not classified as homonyms, the following words are also frequently confused and misused. (These are also discussed in the Handbook.)

about, around
adapt, adopt
advice, advise
affinity, aptitude
aggravate, annoy
allergic, allergenic
allusion, delusion, illusion
alternate, alternative
among, between
amount, number
another, additional
anxious, eager
appraise, apprise
balance, remainder
biannual, biennial, bimonthly, biweekly
breadth, breath
continual, continuous
damage, damages
defective, deficient
diagnosis, prognosis
dilemma, problem
duplicity, duplication
eminent, imminent
entomology, etymology
fewer, less
forced, forceful, forcible
fortuitous, fortunate
full, fulsome
healthful, healthy
imply, infer
lay, lie
less, fewer
loose, lose
luxuriant, luxurious
majority, plurality
mutual, common
people, persons
percent, percentage
prescribe, proscribe
raise, rise, raze
recurrence, reoccurrence
repairable, reparable
vice, vise

See also **Malapropisms.**

HYPHEN. The most common use for a hyphen is to divide words at the ends of lines. However, the hyphen is also used:

1. To indicate a compound modifier (adjective): *one-piece* molding, *first-class* fare, *300-horsepower* motor. The use of hyphens in compound modifiers has long been controversial: some writers insist on them, others resist them, and still others are not certain when to use them.
   A simple test to determine whether a hyphen is needed in a com-

pound modifier: try each modifier without the other. If each sensibly modifies the noun independently, the hyphen is unnecessary. In the first of the preceding examples, neither *one* nor *piece* logically modifies *molding;* therefore, the hyphen is needed to indicate a *one-piece* molding.

However, a hyphen is not used in compounds formed with an adverb and an adjective (as a *more effective* method, a *most daring* attempt) or an adverb ending in *ly* and a verbal (a *clearly stated* objective). Compounds such as *well deserved* are hyphenated when they precede a noun (*well-deserved* reward) but are written without a hyphen when used in the predicate. (The reward was *well deserved.*)

2. To prevent misinterpretation or misreading of other compound words.
   a. *Compound nouns,* consisting of two nouns, a verb plus a noun, or a single letter plus a noun:

      starter-generator A-frame
      receiver-transmitter I-beam
      cure-all T-square

   b. *Compound verbs:*

      heat-treat temperature-condition

   c. *Compound adverbs:*
      (1) Secure the bolts *finger-tight.*
      (2) He was forced to land *cross-wind.*

3. To prevent confusion with another word:

   re-cover (cover again) BUT recover (regain)
   re-collect (collect again) BUT recollect (remember)
   re-form (form again) BUT reform (improve behavior)

4. To prevent ambiguity in compounds such as a light blue coat—a coat that is light blue or a blue coat that is light? The first is a *light-blue* coat; the second is a *light, blue* coat.
5. To prevent confusion—and an unintentional humorous effect—when a compound begins with a noun in the possessive case.

   camel's-hair brush crow's-foot wrench

6. With a number written as two words (*fifty-four*) or to clarify a compound or mixed number, or a compound unit of measure.

twenty-five	ampere-hour
two and one-third	acre-foot
two-thirds full	foot-pound
1-¾ inches	volt-ampere

7. With two or more adjectives that combine with a following single word to modify a noun. "The first-, second-, and third-stage motors will be tested tomorrow." When used in that construction, the hyphens are called *suspension hyphens*.

   Note: For a complete discussion of the use of hyphens with common prefixes and suffixes, see **Prefixes and Suffixes.**

I

IBID. Ibid. is used in footnotes to indicate that a reference is the same as the immediately preceding reference. See **Documentation,** *Footnotes.*

I.E. An abbreviation for the Latin *id est,* which means *that is.* For its correct usage, see **Abbreviations.**

IF AND WHEN. This is an illogical construction; each of those words has a different connotation. *If* implies a condition; *when* suggests that sooner or later something will take place. For instance, "If and when he finishes his homework, we can go to the movies." That sentence really has two connotations:

1. If he finishes, we'll go; if not, we won't.
2. We're going as soon as he finishes his homework.

That brief analysis should demonstrate that *if* and *when* do not logically belong together.

IMPERATIVE MOOD. See **Mood.**

IMPLY, INFER. Despite the loose use of *infer* to mean *imply,* the former means to draw an inference or to conclude, whereas imply means to suggest or express indirectly. Simply remember that to *infer* is to draw an inference from something *implied* or suggested. For instance:

1. I *inferred* from the article that Mary Brown is a brilliant scientist.
2. The article *implied* that Mary Brown is a brilliant scientist.

INCOMPLETE COMPARISON. See **Comparative and Superlative of Adjectives and Adverbs.**

INDEX. An index is an alphabetical listing of all the topics found in a book, in far greater detail than is found in the table of contents. (See the index at the back of this book.) Indexing is beyond the scope of this volume, however. If you are interested in the process, see the University of Chicago *A Manual of Style* (12th ed., 1969), pp. 405–430.

INDICATIVE MOOD. See **Mood.**

INDIRECT OBJECT. The indirect object of a verb names the recipient of a gift, compliment, information, etc. Sometimes the indirect object is written as the object of a preposition; sometimes it is merely implied by the construction. For instance:

1. He gave the *book* (direct object) to *Helen* (indirect object).
2. He wrote the *book* (direct object) for *Mary* (indirect object).
3. He gave *Helen* (indirect object) the *book* (direct object).
4. He contributed *$100* (direct object) to the *scholarship fund* (indirect object).

INDIRECT QUOTATION. An indirect quotation is a written or spoken summary of someone else's statement. It is summarized in different words; thus no quotation marks are used:

Indirect: The manager said that sales were picking up.
Direct: The manager said, "Sales are increasing."

INFINITIVE. See **Split Infinitive.**

INFLATED EXPRESSIONS. How many variations of the word *because* can you think of? Two? Three? Are you ready for seven?

due to the fact that
by virtue of the fact that
owing to the fact that
for the reason that
in view of the fact that
on the grounds that
on account of the fact that

Now add to those the inflated forms for *if* (in the event that) and *although* (in spite of the fact that), and you'll have nine shining examples of how words are wasted.

How often do you use the following inflated phrases?

prior to
subsequent to
in the course of
the greatest percentage of
in the neighborhood of
during the time that
most of the time
a large number of
on the occasion of

Now simplify those phrases to *before, after, during, most, about, while, frequently, many,* and *on.* Would you agree that those one-word synonyms are identical with the blown-up phrases?

Next, examine these verb-noun combinations:

take into consideration
leave out of consideration
make an approximation
put on a demonstration
effect the standardization of
come to a conclusion

But why use a verb and a noun when the verb says it all: *consider, disregard, estimate, demonstrate, standardize,* and *conclude.*

And of course you recognize these well-known prepositional phrases:

green in color
neat in appearance
rectangular in shape
controversial in nature
a charge in the amount of $3.95
twenty-four in number
a period of two months
spring of the year

How else could it be green but in color? or neat? or rectangular? or controversial? The charge is clearly an amount, twenty-four is definitely a number, two months is certainly a period, and spring has always been part of the year. So why say the same thing twice?

Still other types of inflated prose can be created with these common usages:

1. The unnecessary use of relative modifiers (introduced by who, which, and that):

   Inflated: Wages *that are paid to* line workers account for 70 percent of the payroll.
   Simplified: Line workers' wages account for . . .

   Inflated: A graduate *who has had* some laboratory experience will qualify.
   Simplified: A graduate with some laboratory experience will qualify.

Inflated: Water *that is high in* mineral content
Simplified: Highly mineralized water

2. The excessive use of *it is* and *there are.*

   Inflated: It is the belief of most athletes that confidence is important.
   Concise: Most athletes believe that . . .

   Inflated: There are many theories that explain supply and demand.
   Concise: Many theories explain supply and demand.

3. The excessive use of meaningless qualifiers (as in the following expressions): the *very* best, *most* satisfactory, *extremely* precarious.
4. The excessive use of **prepositional phrases.** See **Prepositional Phrases.**

In part 1 (p. 18), I mentioned how most of us are given to circumlocution in conversation. Just this morning my radio informed me that "temperaturewise, the day will be on the cool side." On other occasions I have been advised (radiowise) that "it will be rather warm today as far as temperatures are concerned." Another broadcast informed me that "as far as prices are concerned, inflation seems to be here to stay."

Those three statements are excellent examples of redundancy and indirect expression. The first simply means, *it will be cool today;* the second means, *it will be warm today.* The third statement defies analysis. (Go ahead, analyze it.)

Synonyms for circumlocution include *periphrasis, pleonasm, tautology, prolixity, diffuseness, verbiage,* and *verbosity,* all of which denote an excess of words.

Periphrasis is the substitution of roundabout phrases, often prepositional phrases, for a more direct expression. For instance, "this car of mine" instead of "my car"; "a book of large proportions" instead of "a large book"; "a person with a gift for athletics" instead of "a gifted athlete."

Pleonasm and tautology repeat the obvious: "I listened to him *as he talked*"; "I saw it *with my own two eyes*"; "a house *located* in the valley"; "a *new* innovation"; "*past* history"; "my *personal* belief"; "red *in color*"; "*true* facts"; "*end* result or product."

Prolixity and diffuseness go into endless details and so tend to hide the main idea being discussed. Verbiage and verbosity denote an excess of useless words that add little to the ideas being discussed.

See also *Watch for Redundancy and Circumlocution* in part 1, p. 17.

INFRA (as a prefix). See **Prefixes and Suffixes.**

IN-LAW (as a suffix). See **Prefixes and Suffixes.**

INNOVATION. An innovation is a new thing or idea. Don't call it a *new* innovation.

INSTRUCTIONS. See **Process Explanation.**

INTERJECTION. An interjection is an emphatic expression usually followed by an exclamation mark. See **Exclamation Mark.**

IRREGULAR VERB. See **Verb.**

IT. *It* is the third-person impersonal singular pronoun. *It* is frequently useful as a pronoun and necessary in such constructions as *it is raining, it can't be proven.* However, *it* is frequently and unfortunately used in such indirect expressions as *it is estimated, it is my belief, it is suggested.* In such constructions, *it* becomes the grammatical subject of the sentence, whereas the actual subject falls into the predicate, often as the object of a preposition. For a more complete discussion of indirect and weakened locution, see **Expletive; Inflated Expressions.**

ITALIC TYPE, ITALICIZING. Italic type in printed copy is slanted to the right and often printed in a light "face" to give it a distinct emphasis. Printers generally refer to ordinary text type that is not italicized as *roman.* The use of italic type in printing corresponds to underlining on the typewriter or in handwritten copy.

The principal uses for italic type are to designate (1) titles of books, magazines, newspapers, reports, etc.; (2) foreign words; and (3) words to be read as "words" (rather than for meaning), e.g., "Don't misuse the word *disinterested.*" Italics are sometimes used to lend special emphasis to words and phrases. In typewritten and handwritten copy, italics are indicated by underlining. For examples of the main uses for italic type and underlining, see **Underlining.**

ITS, IT'S. *Its* is the possessive form of the impersonal pronoun *it* and requires no apostrophe. *It's* is a contraction that means *it is.* For example:

1. You can't judge a book by *its* cover.
2. *It's* (it is) too early to tell.

J

JARGON. The specialized language used by members of a trade or profession is called *jargon,* or *shoptalk.* Lumber workers, for example, speak of *chokers, dogs, green chain,* and many other terms not generally recognized by those outside the lumber trade. For a discussion of the communication problem caused by a writer's use of jargon, see **Technical Terms and Jargon.**

L

LATITUDE AND LONGITUDE. With expressions of latitude and longitude, the abbreviations for degrees, minutes, and seconds are used. For instance, 42°21′16″ N. See also **Numbers.**

LAY, LIE. *Lay* is transitive and thus takes an object. *Lie* is intransitive. The principal parts:

		Participles	
Present	**Past**	**Present**	**Past**
lay (transitive)	laid	laying	laid
lie (intransitive)	lay	lying	lain

Here are some examples of correct usage:

1. *Lay*
   a. I *laid* (not lay or layed) the book on the table.
   b. The table was *laid* (not layed) for six.
   c. The hen has *laid* (not layed or lain) an egg a day for the past year.
   d. They are *laying* the foundation today.
2. *Lie*
   a. He often *lies* (not lays) down before dinner.
   b. I was *lying* (not laying) down when the earthquake struck.
   c. The house *lies* (not lays) between the woods and the river.
   d. He was fired for *lying* (not laying) down on the job.

To sum up, you *lie* down to rest; you *lay* a book on the table. See also **Verb,** *Regular and Irregular Verbs.*

LEAD, LED. The confusion arises from the use of *lead* as both a noun (metal) and a verb (to lead the parade). *Led,* however, is the past tense and past participle of *lead* (verb):

1. He will *lead* the parade tomorrow.
2. He *led* the parade yesterday. He has *led* many parades.
3. *Lead* is used in a variety of products.

LESS, FEWER. See **Fewer, Less.**

LETTER OF APPLICATION. See **Application Letters and Résumés.**

LETTER REPORTS. See **Reports.**

LIABLE, LIKELY. Although these two are interchanged in conversation, *liable,* in the strictest sense, means *legally obligated or susceptible.* In technical writing, the better practice is to use *liable* in that literal meaning and to use *likely* to mean *probable.*

1. He was found *liable* for the damage to his neighbor's fence.
2. It's *likely* to start raining very soon.

LIBRARY RESEARCH. Technical and scientific writers, and particularly college students, should become acquainted with the library and know how to perform research. Let's explore the conventional methods of library research and then look briefly at computerized information networks to which many libraries are connected.

Conventional Library Research

To research any topic, you could begin with an encyclopedia or other general reference—not for specific information but to locate specific sources on your topic. Most broad subjects in encyclopedias include bibliographies, which can direct you to specific sources. Here are four general encyclopedias you will find useful:

Encyclopaedia Britannica
Encyclopedia Americana
Collier's Encyclopedia
Columbia Encyclopedia (one volume)

Many specialized reference works are also available. Of particular interest to those doing scientific research are:

Bibliography of North American Geology

Encyclopedia of the Biological Sciences

McGraw-Hill Encyclopedia of Science and Technology

Van Nostrand Reinhold Encyclopedia of Chemistry

Van Nostrand Reinhold Encyclopedia of Physics

Van Nostrand's Scientific Encyclopedia

For those interested in the social sciences, here are five standard reference works:

Business Information: How to Find and Use It

Encyclopedia of American History

Encyclopedia of Management

Encyclopedia of Psychology

Encyclopedia of World History

Those are only a few of the many specialized references available in most libraries; a complete list of reference works might run to more than a dozen pages. The best way to find those helpful to you is to visit your library. In an hour or so, you will be able to find out exactly what is at hand in the field you want to investigate.

You will also want to acquaint yourself with the indexes to periodicals. The contents of more than one hundred general magazines and journals (since 1900) are listed in the *Readers' Guide to Periodical Literature*. Articles dated before 1907 are listed in *Poole's Index to Periodical Literature*. Other indexes list articles in specialized or technical journals. A few of those are:

Applied Science and Technology Index

Business Periodicals Index

Education Index

Humanities Index

Social Science Index

The introductory pages in a periodical index list the magazines and journals shown in that index, a key to abbreviations, and other information. By consulting the front pages, you can usually determine whether a particular index will help you.

The card catalog. All of the library's holdings are listed in the card catalog, a complete alphabetical index of everything available. Your first step is to learn the arrangement of the catalog. In most libraries, the same book is listed by (1) author's last name, (2) title (omitting *A, An,* or *The*), and (3) subject (see sample cards). All three may be combined in one continuous alphabetical file or the file may be divided. In the California State Library in Sacramento, for instance, authors and titles are in one file, subjects are in a second file, and periodicals are in a third file. (In this particular library, there is also a separate "California" file.) You can quickly determine the arrangement in your library; if in doubt, ask the librarian.

A working bibliography. To expedite your search for information, you will find it useful to compile a preliminary list of sources—more than you

SUBJECT CARD

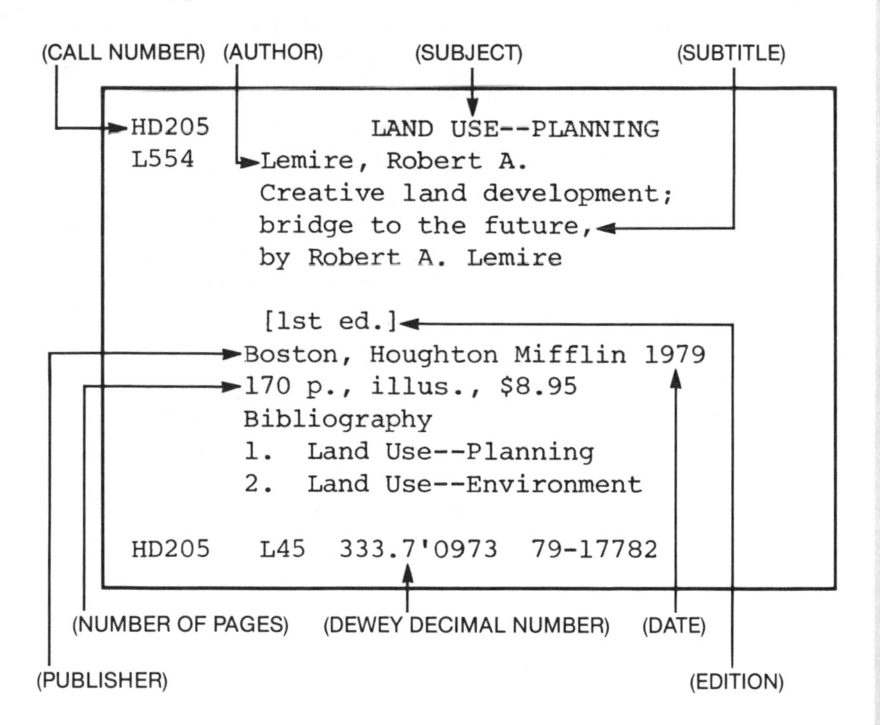

L

TITLE CARD

```
HD205    Creative land development
L554     Lemire, Robert A.
```

AUTHOR CARD

```
HD205    Lemire, Robert A.
L554     Creative land development
```

actually need. Let's assume you are interested in a subject that began making news in the 1970s: the application of lasers to medical practice. In the *McGraw-Hill Encyclopedia of Science and Technology,* you would find specific applications of lasers to many fields, including medicine, and a bibliography of twenty-four sources. In addition, you could find fifteen references in the *Encyclopaedia Britannica,* nine in the *Encyclopedia Americana,* four in the *Van Nostrand Reinhold Encyclopedia of Physics,* and, undoubtedly, several in the subject file of the card catalog.

You will also want to examine the *Readers' Guide to Periodical Literature* for the titles of articles in magazines and journals that bear directly on your topic. In this instance you will find them under *Lasers—Medical Use.* Perusing the 1979 and 1980 issues of the *Readers' Guide,* I found fourteen references to the medical use of lasers in the 1979 issues; in the 1980 issues, through July alone, there are five.

Here is an almost new technology, and already an extensive bibliography has been compiled. This should suggest that research material is available on almost any subject you might choose to investigate.

Taking notes. An efficient system is to take notes on three-by-five cards, limiting yourself to a single subject on each card; you will find it much easier to arrange the cards if you stay with one subject on each. Write the call number in the upper left-hand corner; then follow with the author, title, and publication date. Be sure to include all the information shown on the following sample cards. You will then have everything you need to compile a bibliography.

BIBLIOGRAPHY CARD FOR A BOOK

call no.	HD205-L554
author	Lemire, Robert A.
title	Creative Land Development, 1st ed., 1979
publisher	Houghton Mifflin, Boston

BIBLIOGRAPHY CARD FOR A PERIODICAL

call no.	UC 749-G4-01
author	Seltz-Petrash, Ann
title	Marketplace Solutions to Air Pollution
publisher	Civil Engineering, Vol. 50, No. 1 January 1980, p. 68

L

Avoiding plagiarism. When taking notes, you may summarize or paraphrase the author's ideas, or you may wish to quote an author verbatim. When you do quote, be sure to use quotation marks to distinguish between your words and those of the author. To use the author's words—or even a paraphrase of the author's words—as your own constitutes plagiarism. Of course, you may use the author's words or ideas as long as you do not represent them as your own.

To guard against unintentional plagiarism, remember that whenever you use the words or ideas of another, you must document your source. Accordingly, if you merely rearrange a few words in an original passage, or include a few words of your own, and then pass the idea along as yours, you have committed plagiarism. Plagiarism is not only illegal but also immoral. It is grounds for dismissal from many colleges and universities.

Here, for example, is a sentence from page 32 of *An Economic Interpretation of the Constitution of the United States*, by Charles A. Beard:

Creditors, naturally enough, resisted all these schemes in the state legislatures, and failing to find relief there turned to the idea of a national government so constructed as to prevent impairing the obligation of contracts, emitting paper money, and otherwise benefiting debtors.

Now, here is a paraphrase of that passage:

It was only natural that creditors should resist those schemes of the state legislatures, which were largely dominated by the agricultural and nonpropertied classes. Since the property interests could obtain no relief at the state level, they turned to the idea of a federal government that would (1) forbid the states from issuing paper money, (2) prevent laws that would impair the obligation of contracts, and (3) otherwise protect the interests of property owners.

The paraphrase constitutes plagiarism, whether unconscious or intentional. It obviously asserts the ideas of the original author and thus must be attributed to him. This can be done by accrediting his assertion in the paraphrase or by footnoting the source. For instance:

It was only natural, *Beard asserts,* that creditors . . . He further asserts that since property owners could obtain no relief . . . etc.

That use of the author's ideas is perfectly acceptable. Whenever you are in doubt about whether you are plagiarizing the ideas of another, be sure to credit your source.

Computer Output Microfilm Catalog (COM CAT). An innovation in library cataloging is the computer output microfilm catalog (COM CAT), which is beginning to supplant the conventional card catalog in many libraries. COM CAT, a microfilm reader approximately the size of a portable television receiver, contains author, title, and subject files, which are projected on a lighted screen. You can rotate the files by merely pushing a button or move them manually with a hand control.

A great advantage of COM CAT is that you can find all the works by an author, or all the books on a particular subject, in one continuous list. The files contain the same information found in the card catalog. And, if you live in a city with a number of branch libraries, COM CAT will also tell you in which libraries each book is available.

Computerized Information Networks

You should also become acquainted with the computerized information networks to which many libraries, college libraries in particular, have subscribed. Three such networks provide this information service:

- The Research Library Information Network (RLIN) at Stanford University, Palo Alto, California
- The Ohio College Library Center (OCLC) in Columbus, Ohio
- The Washington Library Network (WLN) in Washington, D.C.

Right now, the services provided by the networks are principally for their member libraries, but there are benefits for library patrons as well. For instance, if you were searching for a rare book, a member library could inquire, through the network, whether the book is available in another library. If the book is located, the member library can usually arrange an interlibrary loan.

Member libraries can also use the network service to verify bibliographic sources. In addition, the networks provide cataloging service, including catalog cards; Library of Congress MARC (machine readable cataloging) records; and magnetic tapes of subscribers' records.

Briefly stated, the purpose of the network service is to:

- improve the services offered by member libraries;
- furnish library staffs with hard-to-find information;
- furnish personalized service to library patrons.

To find out whether your library is a member of one of the networks, and to learn what it can mean to you, ask your librarian for more information.

LIKE (as a suffix). See **Prefixes and Suffixes.**

LIKE, AS. The confusion between these two words inspired the former well-known TV commercial that began, "Winston tastes good like a cigarette should." The announcer then explained that the use of like was "improper grammar" and changed the line to ". . . *as* a cigarette should."

Like is most often a preposition and should not be used as a conjunction—at least in formal writing. Simply remember that *like* means *similar to* or *desirous of:*

1. Winston tastes *like* (similar to) a cigarette.
2. Winston tastes good *as* (conjunction to introduce a subordinate clause) a cigarette should.
3. It looks *like* (indicative of) a good day for fishing.
4. I feel *like* (desirous of) going fishing.

Like is sometimes used as an adjective meaning similar, as in the construction, "this and *like* events."

LINKING VERBS. See **Verb.**

LISTING TECHNIQUE. See **Enumeration.**

LOOSE, LOSE. When you mislay something, you *lose* it. When you let a bird out of a cage, you turn it *loose.*

LOOSE SENTENCE. See **Sentence Types and Construction.**

LUXURIANT, LUXURIOUS. The first word of this pair means growing abundantly or displaying abundance, as a luxuriant beard, lawn, or garden. *Luxurious* means displaying luxury, as a luxurious home or automobile.

M

MAJORITY, PLURALITY. *Majority* is considered singular when it refers to a definite number; it is considered plural when it refers to a group of individuals.

Singular: The senator's majority was 5,651 votes.
Plural: The majority were in favor of the proposition.

Plurality refers to the number by which a winning candidate defeats two or more opponents. For instance, if three candidates ran for one office and the votes were tallied as 4,000, 3,400, and 3,000, the winner would have a *plurality* of 600 votes.

MALAPROPISMS. Not long ago I read a newspaper story describing the prerequisites [sic] to which United States senators are entitled. This may have caused some confusion for readers unacquainted with *perquisites,* which are the privileges that go along with high political office—paid staff help, free postage, etc. Obviously, somebody at the newspaper was confused, and the wrong word appeared in the story.

Much the same thing happens every day to writers who are unaware of the distinction between such homonyms as *affect* and *effect, dual* and *duel, capital* and *capitol, principal* and *principle.*

A confusion between two words that resemble each other—either in spelling or in pronunciation—is called a *malapropism,* after the humorous character Mrs. Malaprop in Richard Sheridan's play *The Rivals.* In this case, the eighteenth-century playwright deliberately used incorrect words for their humorous effect, little suspecting that his efforts would add a new word to the English language.

Intentionally using malapropisms to create humor is one thing. I do believe, however, that some sort of award should have gone to the community official who, some years ago, presented a local doctor with a certificate honoring him for his *meretricious* [sic] service to humanity.

In part 1, p. 16, I mentioned several instances where writers had created malapropisms without even trying. Unfortunately, writers can use incorrect words all too easily, unless they make a conscientious effort to learn the difference between words that resemble each other.

The usual problem created by malapropisms is more one of a writer's image than of a reader's confusion. If a writer slips and readers catch her, they may decide she has only a superficial knowledge of her subject. Unfortunately, readers may remember a writer's errors long after they have forgotten the rest of the message.

A word to the wise: when in doubt, look it up!

See also **Homonyms; Spelling.**

MANTEL, MANTLE. The facing around the fireplace is the *mantel;* the shelf above it is the *mantelpiece.* The *mantle* is the light covering of snow on the ground or the layer of earth between the crust and the core.

MEMORANDUM REPORTS. See **Reports.**

MEMORANDUMS. A memorandum is really an "internal" letter, often addressed to a group or to a number of individuals. Those who work for large organizations are no doubt familiar with "all-employee" memorandums. A memo is prepared on a preprinted form with a subject line, which relieves the writer of any need for a long, rambling opening. There is no need to begin a memo with, "The purpose of this memorandum is . . . etc." The communication is obviously a memorandum (it says so at the top), and its purpose is stated in the subject line. Thus, writers can be specific from the opening word. For example:

Memorandum Showing Preprinted Heading and Subject Line

```
                    M E M O R A N D U M

  TO:  All Program Managers          DATE:  June 4, 1981
FROM:  Glenn T. Jackson           SUBJECT:  Distribution of
       Assistant Director         reports to the Legislature

Our present procedure is to send copies of all reports to
the secretary of the Senate and the chief clerk of the
Assembly.  At their request, the director is now asking
that those officials receive only reports required by the
Water Code and those prepared in response to legislative
resolutions.

To assist Mrs. Thompson, who is responsible for distribution,
I would like program managers to identify reports that should
be sent to the secretary and the chief clerk.  (We will con-
tinue to send copies of all reports to the Assembly Water
Committee and the Senate Committee on Water Resources.)  When
submitting a report for approval to print, program managers
should indicate whether the report is (a) required under the
Water Code, (b) in response to a resolution, or (c) discre-
tionary.

Please begin this new procedure immediately.
```

METRIC SYSTEM. The word *metric* was adapted from the Greek word *metron,* which means a measure. Another derivation of the same word is meter (or *metre*), which denotes the unit of length in the metric system. Recently the metric system has been updated and is now officially called the *International System of Units* (abbreviated as SI). According to plan, we may eventually be using it in our daily lives. If you have studied any of the natural sciences, you are undoubtedly familiar with a number of metric terms.

Although at first glance the metric units may seem strange, the SI system is actually less cumbersome than the United States system of weights and measures. Look, for example, at our common units of mass (weight)—ounces, pounds, and tons. In the SI system, the different sizes of the base unit (the kilogram) can be calculated quickly by simply shifting a decimal point. Thus the metric, or SI, system is a base-10, or decimal, system.

The base unit of length, for example, is the meter. Smaller units include the millimeter (1/1000 of a meter), the centimeter (1/100), and the decimeter (1/10). Larger units include the dekameter (10 meters), the hectometer (100 meters), and the kilometer (1,000 meters). The prefix for one millionth (*micro*) is not used with the meter, because the word *micrometer* also denotes a measuring instrument. Instead, the term *micron* is used for this particular unit.

The SI system, which was established by international agreement, is constructed on a foundation of seven base units plus two supplementary units; all other SI units can be derived as powers of 10 of the following base units:

1. the meter (or *metre*), the unit of length;
2. the kilogram, the unit of mass (weight);
3. the second, the unit of time;
4. the ampere, the unit of electrical current;
5. the candela, the unit of luminous intensity;
6. the mole, the unit of amount of substance;
7. the Kelvin, the unit of temperature. The Kelvin, however, will undoubtedly be confined to the laboratory. The common scale for temperature will be the Celsius (also called centigrade).

In our daily lives, most of us will be concerned with units of length, mass, and temperature only; the other four are identical in both the U.S. and SI systems anyway.

Converting from one system to another will require some practice, but it can be done rather easily with a conversion chart and a calculator. Use the table of common conversion factors to interchange U.S. and SI units quickly. Simply follow the instructions in the table heading to convert U.S. units to SI units and vice versa.

Common Conversions (accurate to six significant figures)

Symbol	*When you know*	*Multiply by*	*To find*	*Symbol*
		U.S. system to SI system		
in.	inches	25.4	millimeters	mm
ft	feet	0.3048	meters	m
yd	yards	0.9144	meters	m

M

Common Conversions (cont.)

Symbol	*When you know*	*Multiply by*	*To find*	*Symbol*
mi	miles	1.609 340	kilometers	km
yd^2	square yards	0.836 127	square meters	m^2
yd^3	cubic yards	0.764 555	cubic meters	m^3
(none)	acres	0.404 686	hectares	ha
qt	quart	0.946 353	liters	l
oz	ounces (avdp)	28.349	grams	g
lb	pounds (avdp)	0.453 592	kilograms	kg
°F	degrees, Fahrenheit	5/9 (after subtracting 32)	degrees, Celsius	°C
		SI system to U.S. system		
mm	millimeters	0.039 370	inches	in.
m	meters	3.280 840	feet	ft
m	meters	1.093 610	yards	yd
km	kilometers	0.621 371	miles	mi
m^2	square meters	1.195 990	square yards	yd^2
m^3	cubic meters	1.308 950	cubic yards	yd^3
ha	hectares	2.471 050	acres	(none)
l	liters	1.056 690	quarts	qt
g	grams	0.035 274	ounces (avdp)	oz
kg	kilograms	2.204 620	pounds (avdp)	lb
°C	degrees, Celsius	9/5 (then add 32)	degrees, Fahrenheit	°F

As you can see from the chart, 1 inch is about 25 millimeters or 2.5 centimeters. One meter is slightly longer than a yard, and one mile is a bit more than 1.6 kilometers. One liter is about 5 percent more than a quart, and 1 gallon equals about 3.75 liters.

The chart is quite simple to use. For example, if you drive 10 miles to work or school, you will drive 10 times 1.609 or roughly 16 kilometers. The legal speed limit of 55 mph, by the way, will become about 88 kph (kilometers per hour).

If you buy 2 pounds of your favorite coffee, you'll come home with not quite 1 kilogram (0.9 kilogram). If you also buy a 1-liter carton of milk, you'll find that it holds slightly more than the quart you used to buy (1.05 quarts, or about 5 percent more).

If you weigh 150 pounds, you will also weigh in at 68 kilograms (150 times 0.453). And in the summertime, when it's 90°F, you'll have to get used to swimming in 32°C weather (90 − 32) times 5/9 = 32.

Now, here are the multiples and prefixes that can be applied to all SI units:

Multiples and submultiples	*Prefix*	*Symbol*
1 000 000 000 000 = 10^{12}	tera	T
1 000 000 000 = 10^{9}	giga	G
1 000 000 = 10^{6}	mega	M
1 000 = 10^{3}	kilo	K
100 = 10^{2}	hecto	h
10 = 10^{1}	deka	da
base unit 1 = 10^{0}	——	——
0.1 = 10^{-1}	deci	d
0.01 = 10^{-2}	centi	c
0.001 = 10^{-3}	milli	m
0.000 001 = 10^{-6}	micro[1]	μ
0.000 000 001 = 10^{-9}	nano	n
0.000 000 000 001 = 10^{-12}	pico	p

[1] *Micro* is not used with the *meter*. This unit is called a *micron*.

Finally, in the SI system, no commas are used between the digits of multiple numbers. Instead a space is left: 1 000; 13 461; 1 976 457; 3 657 413 265.

MICRO (as a prefix). See **Prefixes and Suffixes.**

MID (as a prefix). See **Prefixes and Suffixes.**

MINI (as a prefix). See **Prefixes and Suffixes.**

MISPLACED MODIFIERS. See *Be Careful with Modifiers*, part 1, p. 20; and **Dangling Constructions** in the Handbook.

MODIFIERS. Modifiers are words, phrases, and clauses that affect the meanings of other words, phrases, and clauses. A modifier may be a(n):

Adjective: a *blue* car

Adverb: John ran *quickly*

Noun: a *laboratory* report

Verb: a *running* track

Phrase: After the game, they all went home.

Clause: If sales increase, we will need more help.

See *Be Careful with Modifiers*, part 1, p. 20. Related Handbook articles include **Adjectives; Adverbs; Clauses and Phrases; Comparative and Superlative of Adjectives and Adverbs; Dangling Constructions; Restrictive and Nonrestrictive Modifiers.**

MONO (as a prefix). See **Prefixes and Suffixes.**

MOOD. The mood of a verb shows the attitude of a writer or speaker about the action or condition expressed:

1. The *indicative,* or *declarative,* mood states a fact or asks a question:

   a. What is the date today?

   b. Today is Friday, February 28, 1980.

2. The *imperative* mood gives a command or makes a request:

   a. Close the door.

   b. If I can help, please call me.

3. Today, the *subjunctive* mood is used principally to indicate a condition contrary to fact, a doubt, or a wish. A "contrary" situation or a wish is usually shown by the use of *were* where *was* is customary; a doubt is often shown by the use of *could* or *might.*

The subjunctive mood has all but vanished from today's comparatively terse prose. Some years ago, when writing was more formal, the subjunctive was more prevalent. For example:

Formal: We shall use force, if that *be* necessary.
Terse: If necessary, we will use force.

Here are examples of the subjunctive mood as used today:

Contrary to fact

a. If the sun *were* to stop shining, all life would disappear.
b. He talks as though he *were* the manager.

Note: Factual statements do not require the subjunctive: "If you knew he *was* wrong, you should have said so."

Wish

a. I wish I *were* a better carpenter.
b. If only John *were* here.

Doubt

a. He *might* come, but don't count on it.
b. Do you think we *could* finish in two weeks?

N

MORE, MOST. See **Comparative and Superlative of Adjectives and Adverbs.**

MS. (as a title for a woman). See **Correspondence.**

MULTI (as a prefix). See **Prefixes and Suffixes.**

MUTUAL, COMMON. *Mutual* refers to the relationship between two persons, as a mutual agreement or goal. When the reference is to a group, however, the correct adjective is *common.* For example, "The team shares a common goal."

N

NATURE. A well-known, overused bit of deadwood, the meaning of which is almost always expressed by the adjective preceding it:

If your writing problems are minor *in nature,* you'll have little need for this handbook.

See also **Inflated Expressions.**

NEITHER . . . NOR. See **Correlative Conjunctions.**

NOMINATIVE CASE. See **Case.**

NON (as a prefix). See **Prefixes and Suffixes.**

NONE. *None* (pronoun) may be used to mean *not one* or *not any.* In the first sense, *none* requires a singular verb; in the second sense, the verb should be plural. For example:

Singular: I bought three books, none (*not one*) of which was very expensive.
Plural: None (*not any*) of them have finished their work.

NONRESTRICTIVE MODIFIERS. See **Restrictive and Nonrestrictive Modifiers.**

NON SEQUITUR. A *non sequitur* is an illogical statement, really an inference or a conclusion that does not follow from the evidence presented as a basis for the statement. For example:

Because I own an automobile and drive it on public roads, the law requires all drivers to carry automobile insurance.

That statement would undoubtedly disturb many readers until they had determined the writer's actual meaning:

Because I own an automobile and drive it on public roads, I am required by law to carry automobile insurance.

Now the statement is logical; the second clause logically evolves from the opening clause. Here is another example:

When the engine fails, the helicopter transmission is designed to permit the main rotor to turn in the direction of rotation.

Again an illogical statement—the transmission permits the main rotor to turn in the direction of rotation at all times, not just when the engine fails. As revised:

The transmission is designed to permit the main rotor to continue turning in the direction of rotation, even if the engine should fail.

Read your sentences over carefully to make certain they evolve logically from previous clauses or sentences.

NOT ONLY . . . BUT ALSO. See **Correlative Conjunctions.**

NOUN.

1. Nouns name persons, places, objects, and qualities or actions, such as virtue, whiteness, laughter. The names of common inanimate objects are called *common* nouns; names of persons and specific places and institutions are called *proper* nouns and are always capitalized (see **Capitalization**). Of course, the same noun can be common or proper, depending on the context:

   Common: He works for a *bank.*

   Proper: He works for the *Citizens National Bank.*

2. Other parts of speech may also be used as nouns, notably the present participial form of **verbs** (a verb plus *ing*): swimming, walking, being, etc. This particular form is also called a *gerund,* or *verbal noun.* Even an **adjective** may be used as a noun: "*Red* is my favorite color."
3. A special type of noun writers should treat with care is the *collective* noun, whose singular form names a group of items or persons. Some common collective nouns:

audience	corporation	gang	number
band	crowd	group	public
class	department	herd	remainder
club	electorate	jury	team
committee	flock	majority	throng

   Three general rules for the use of collective nouns:

   a. *A collective noun must agree (in number) with the verb and the pronoun (if any).* The rule is simple: when the noun connotes the group as a whole, both verb and pronoun should be singular. Conversely, when the noun connotes the individual members of the group, both the verb and the pronoun are plural. For example:

      Singular:

      (1) The Accounting Department will hold *its* (not their) spring dance at the country club.

      (2) The club is conducting *its* (not their) annual membership drive.

      (3) The number of students studying computer science *is* increasing.

      Plural:

      (1) Despite the regulation against it, a number of students *were* smoking in the laboratory.

N

(2) The audience at the outdoor concert *were* scattered throughout the park; some *were* listening intently, while others *were* apparently bored.

(3) The majority *were* in favor of the new laws.

b. *Expressions denoting quantities are usually treated as singular.*

(1) Ten feet of snow *has* fallen since December 10.

(2) Eight thousand dollars for an automobile *is* a great deal of money.

(3) Two times two *is* four; five plus three *is* eight.

c. *When a collective noun is used in the plural sense, a predicate noun or the object of a verb should also be plural.*

(1) The public, particularly those who depend on their *cars* (not *car*) to get to work, were advised to remain in their *homes* (not *home*) during the storm.

(2) Because of the storm, only 30 percent of the electorate cast their *votes* (not *vote*) today.

(3) The senior class will receive their *grades* (not *grade*) early.

NOUN MODIFIERS. See *Avoid Superconcise Writing*, part 1, p. 19.

NUMBER. Number is a **collective noun** requiring either a singular or a plural verb in accordance with the meaning—total or individual—expressed. A simple rule to help you decide on the number of the verb: when *number* is preceded by *a,* the construction is plural; when preceded by *the,* it is singular:

1. *A* large number of students *have* taken the course.
2. *The* number of successful graduates *is* amazing.
3. *A* number of graduates *were* in the audience.
4. *The* number of players permitted on each team *has* been increased to twenty-five.

NUMBERS. The use of numbers may often lead writers into grief and readers into confusion. Newspapers, magazines, and business and governmental organizations—particularly those that issue reports—usually establish a numerical style, and their writers simply adopt and follow it. For writers "at large," however, the path is not so clear. Yet there are a number of conventions that enable technical writers to express numbers clearly and concisely.

1. If a number can be expressed in no more than two words, write it out: nine, fifteen, twenty-five, one hundred; BUT 127, 6,578, 39,675. Here are two notable exceptions to that rule:
   a. Many organizations, particularly those that issue technical reports, have adopted a "technical" style, in which all numbers greater than *nine* are expressed as figures. Should your instructor specify technical style, simply remember to write out *one* through *nine* and express all larger numbers as figures: *three* books, *eight* cartons, 10 tubes, 39 pencils, 150 envelopes.
   b. Numbers that precede units of measure are usually expressed as figures regardless of size: 2 inches, 7 yards, 6 hours, 5 acres. BUT—
2. Figures are never used to open a sentence. Either write out the number or recast the sentence so that it does not open with a number.
3. Figures may always be used with *prices* ($3.85 or 37¢), *temperatures* (85°F), *time of day* (8:15 a.m.), *degrees of latitude and longitude* (50° 30′ 16″ S), dates (March 3, 1980), and *sports scores* (7 to 4).
4. To write dates, simply use the number for the day of the month without adding *st, rd,* etc.: March 15, 1980, or 15 March 1980 (sometimes used in documentation).
5. In text, do not use figures to represent either ordinal numerals or fractions:
   a. On the *fourth* (not 4th) try, I passed the test.
   b. Almost *two-thirds* (not ⅔) of the students were absent.
6. For large numbers, such as millions and billions, a *mixed form* is used because it is easy to read: *28 million, 103 billion,* etc. Note that this form omits long strings of zeros, which readers will often have to count to determine whether the writer means millions, billions, etc. This form may also be used for monetary expressions. For example, "The contract totaled $350 million (or 350 million dollars)." To summarize: use zeros to express numbers in the thousands (10,000, 340,000). Use the mixed form for numbers in the millions and billions (15 million, 10 billion, 100 billion).
7. Where numbers are used frequently within sentences and paragraphs, readers will have an easier time with figures than with written numbers. For instance:

The crew consisted of 23 carpenters, 14 electricians, 15 plumbers, 7 painters, and 6 general helpers.

8. To express decimal fractions with values of less than 1, precede the decimal point with a zero so that the point will be obvious: 0.3, 0.014, 0.0047.
9. Use a hyphen in mixed numbers: 2-5/8, 3-3/16.
10. When one number immediately precedes another, spell out the small number and express the larger in figures: *four 6-inch pipes, 20 six-inch-long tubes.*
11. Don't write out a number followed by the numerals in parentheses. Although it is sometimes found in contracts, the duplication is unnecessary in ordinary text.

Wrong: The base is three (3) feet long.
Right: The base is 3 feet long.

See also **Metric System.**

O

OBJECT OF A PREPOSITION. See **Preposition.**

OBJECT OF A VERB. See **Direct Object; Indirect Object; Verb.**

OBJECTIVE CASE. See **Case; Pronoun.**

ODD (as a suffix). See **Prefixes and Suffixes.**

ONE OF THOSE . . . WHO. This locution requires a plural verb:

She is *one of those rare persons who require* only four hours of sleep.

Note that the verb agrees with the antecedent of *who* (persons).

ONLY. In conversation, most of us place *only* just before a verb, and we are seldom misunderstood—probably because of the emphasis we place on the word modified by *only:* "I'm *only* going to tell you once." That statement really means, "I'm going to tell you once *only.*" As another example, "I *only* have eyes for you" means, "I have eyes for you *only.*"

In writing, although the meaning will often be clarified by the context, *only* should usually be placed just before or just following the modified word or phrase. Note how the meaning of the following sentences is changed as the position of *only* is shifted:

1. *Only* the teacher can authorize students to work on the research program. (The teacher is the only one who can authorize students to work on the program.)
2. The teacher could assign *only* four students to work on the program. (Only four students could work on the program.)
3. The teacher authorized the students to work on the research program *only*. (The students were authorized to work on the research program and no other program.)
4. The teacher could authorize additional students to work on the program *only* when necessary. (No additional students could be assigned to the program except when necessary.)

Note how the placement of *only* brings out the writer's exact meaning without the additional words found in the explanation of each sentence.

ORAL PRESENTATIONS. Technical writers, engineers, managers, and other professionals may often be called on to make oral reports and briefings—on plans, studies, investigations, and perhaps on their specialties. As a writer, therefore, you should know how to prepare and present an oral discussion. Actually, you can prepare an oral report just as you prepare a written report, i.e., by working up an outline or guide.

Unless you are a skilled speaker, you probably should not try to "read" a written report—word for word, that is. Reading from a prepared manuscript is an art in itself, and nothing can lose an audience's interest more quickly than a poorly "read" paper. Therefore, your presentation will probably be more natural and spontaneous if you rely on notes and a visual aid or two. You can use flip charts on an easel; an opaque projector to project text or artwork on a screen; an overhead projector that projects transparencies, which are clear plastic sheets you can write or draw on; or a slide projector.

Preparation is half the battle in making an oral presentation. If you have prepared your talk well and rehearsed it several times, you will know exactly what you are going to say and do during the actual presentation. Rehearsal is the key to a smooth presentation. If you know what you want to say and do at all times, you can eliminate embarrassing pauses and gaps during the "show."

Let's look at several types of visual aids and then take up the preparation of an actual presentation.

Visual Aids

1. *Flip Charts.* Flip charts are prepared on a large pad that is attached to an easel. You can print the key parts of the presentation on the pad, and then simply flip the pages during your talk. Using a chart is better than trying to write on a blackboard as you are talking. Since you have the material all ready to show, you will not break the continuity of your presentation by stopping to write on the blackboard, nor will you have to turn your back on the audience. Also, the printing on the chart will undoubtedly be much neater and more legible than will hasty scrawls on a blackboard.
2. *Opaque Projector.* With an opaque projector you can show pages of text, illustrations, photographs, etc. Simply arrange the pages you intend to use in correct order, and then insert them as you are talking. Better yet, have someone else show the pages, and you can stand near the screen and use a pointer to emphasize key information.
3. *Overhead Projector.* An overhead projector uses transparencies, which are plastic sheets you prepare in advance. An advantage of the overhead projector is that you can draw on the sheet as it is being projected on the screen. If you are showing a chart, for example, and want to superimpose a line, you can do it while the chart is being shown.
4. *Slide Projector.* If you have slides, you may want them to be the main part of the presentation. As you show each slide you will probably want to comment on it; therefore, you should prepare your comments in advance. You will also need enough light, say at a rostrum, so that you don't become lost in the dark. You can put your comments on three-by-five, or larger, cards and key them to each slide so that you will know what you want to say as each slide is shown.

A Sample Presentation

Assume that you, as a writer for a business firm, have been asked to give an hour-long presentation on "writing better business letters." Now, an hour may seem like an eternity when you are facing an audience, but if you have prepared carefully and rehearsed the presentation, it will probably be over before you know it. The first thing to do is to decide on a procedure.

Naturally, you will want to do more than just talk about letters. You will want to show some examples of effective letters. We'll assume that you have some examples of both good and bad letters that you will show in an opaque projector after you have discussed the principles of modern correspondence. You now decide on this procedure:

1. You will prepare a series of notes on five-by-seven cards as a guide to the presentation.
2. In advance, you will print key points on flip charts, which you can turn during the presentation. On your planning sheets, you decide to draw boxes around those points, representing the charts you will make.
3. After you have presented the important principles of correspondence, you will show several examples of ineffective and effective letters on an opaque projector.

Now, let's map out the entire presentation.

Topic. "Letter-Perfect Correspondence." What? Open with a pun? Remember, if you can get a laugh or two at the beginning, so much the better. If your audience warms up to you immediately, you'll have a better chance of keeping their interest.

Question. Why should we improve our correspondence?

Chart 1

1. To improve our image.
2. To create good will.
3. To get better results with our letters.
4. To take advantage of the personal contact represented by every letter we write.

O

General Principles of Modern Correspondence.

1. Make certain the letter is attractive. "Frame" it like a picture with even margins all around. (Point out that this is the responsibility of a skilled stenographer.)
2. Use standard English, but be as conversational as possible. Avoid these old-fashioned phrases that date back to the horse and buggy days:

Chart 2

OLD: Receipt is hereby acknowledged of yours of 3/12.

NEW: Thank you for your March 12 letter.

OLD: Attached herewith is your check.

NEW: Here is your check.

OLD: As per your request, your policy will be changed.

NEW: As you requested, we are changing your policy.

OLD: Please feel free to contact me.

NEW: Please call me if you have a question.

OLD: Hoping to hear from you soon, I remain

NEW: Please reply by June 10.

3. The reader is all-important. Put him or her "in the picture" right away. Use the YOU attitude to let the reader know you are concerned with his or her problems and needs.

Chart 3

NO: Enclosed is *our* descriptive material on *our* new tax seminar for real estate professionals.

YES: Here is *your* invitation to attend a new tax seminar for professionals like *yourself*.

4. Consider how to say yes and how to say no. If you are granting a request, say so immediately. If you must refuse a request, give logical reasons for your refusal. Then refuse positively.

Chart 4

NEGATIVE: Therefore, we must refuse your request for an adjustment on your account.

POSITIVE: In view of the facts, we believe our charge is correct.

NEGATIVE: We cannot accept merchandise that has been used.

POSITIVE: To protect you and all of our customers, the return of used merchandise is against our policies.

5. If you are granting a partial adjustment or part of a request, give the reader the good news first. Tell what you will do for him or her immediately. Then, try to close positively:

Chart 5

NO: This is the best we can do. (In effect this says, "Take it or leave it.")

YES: I hope this adjustment seems fair to you, Mrs. Johnson. We want you to be satisfied.

6. If you must apologize for a mistake, apologize early in the letter, and then try to close positively. Don't write a "sorry" closing. The reader remembers longest what he or she read last:

Chart 6

NO: Again let me say how sorry I am.

YES: Please try us again, Mrs. Thompson. I am certain you will like our service.

7. If you want the reader to do something by a certain time, say so:

Chart 7

NO: Please return your application as soon as possible. (NOTE: This may mean never!)

Chart 7 (cont.)

YES: Please return your application by May 15.

8. Don't accuse the reader of neglecting or failing to do something:

Chart 8

NO: Since you failed to sign the application, I am returning it for your signature.

YES: Your application wasn't signed, Mrs. Smith. Would you sign it now and return it.

NO: Since you neglected to acknowledge my request for your signature on the policy rider, I am writing again.

YES: Did you receive my request for your signature on the policy rider you asked for?

9. Some general reminders:

Chart 9

1. Be sure your correspondence is letter perfect—no typos or grammatical errors. Especially in brief letters, they will be all too obvious.
2. Use modern English and a friendly tone. Even if the situation is hostile, at least be courteous.
3. Put the reader in the picture immediately. Use the YOU approach to let the reader know you are thinking of his or her needs.

This will end the first part of the presentation. Before you begin showing the letters, you may want to ask for questions. A question period will bridge the gap between the two parts of the presentation and stimulate some audience reaction. In an hour-long presentation, it is a good idea to involve your listeners; in this instance, it will help build interest in the letters you are going to show.

As you present each letter, be prepared to point out its strengths and its weaknesses and ask for suggestions for improving it. It will help if you ask questions that suggest an answer, for example, "Who can suggest a way to put more 'You' into the opening of that letter?"

After you have shown the last letter, give a brief summary of the entire presentation. Keep in mind the old adage about speeches: "Tell them what you're going to tell them, tell them, and then tell them what you have just told them." This will help you wrap things up neatly and concisely. You might have a final chart ready listing the highlights of your talk.

With the sort of plan shown here, some careful preparation, and adequate rehearsal, you should be able to talk on any subject without faltering. Here are a few reminders about oral presentations:

1. Prepare your talk carefully, and rehearse it until you could do it while you are asleep. If you plan to use visual aids, practice with them until you have the timing down. Be sure you know what you are going to do next.
2. Speak slowly, but keep things moving. Avoid deadly pauses, "uh's," "ah's," and "you know's."
3. Be as friendly and conversational as possible. Stay away from "technical" language. No one in the audience will have a dictionary.
4. Explain your topic clearly at the start. Be sure your listeners know what the show is about. Then have a brief summary for the closing.
5. Be prepared for questions and answer them as fully as possible. If you can't answer a question, say so. Never fake an answer. Offer to "look that one up" and get back to the person who asked it.
6. Don't worry about stage fright. It is only natural for anyone facing a group to be nervous. To help overcome tension, keep this in mind: your audience did not come to watch you fail. They came to watch you succeed, and they want you to succeed.

OUTLINE. See *Organization*, part 1, pp. 6–13.

OVER. Compounds with *over* are not hyphenated. See **Prefixes and Suffixes.**

P

PALATE, PALLET, PALETTE. Here is a trio of similar-sounding nouns, each with a different meaning. The *palate* is the roof of the mouth. A *pallet* may denote (1) a portable platform for storing or moving articles, (2) a sleeping mat placed on the floor, (3) an instrument used to mix clay, or (4) a machine part. A *palette* (actually pronounced pallet′ accent on the final syllable) is the board used by an artist to hold oil paints.

PARAGRAPHS. See *Create Meaningful Paragraphs*, part 1, pp. 25–32.

PARALLEL CONSTRUCTION. See **Sentence Types and Construction.**

PARENTHESES. When a complete sentence is placed inside parentheses, the period is placed inside the final parenthesis. Punctuation marks belonging to a sentence of which the parenthetical unit is a part are placed outside the parentheses (including the final period at the end).

PARTICIPLE. A *participle* is a verb form with various functions:

1. Participles are generally classed as *present* or *past,* although they do not indicate definite times by themselves.
   a. A present participle (verb plus *ing*) may be used with an auxiliary (helping) verb to show something "in progress" in the sense of present, past, or future:

      Present: She *is working* in the laboratory this semester. (Working is "in progress" now.)

      Past: When I last saw him, he *was working* in the laboratory. (Working was "in progress.")

      Future: They *will be working* on that project for at least two years. (Working will be "in progress.")

   b. A *past* participle may be used with an auxiliary verb to show continuance or completion (called the *perfect* tense):

      Present: He *has worked* there for twenty years. (He has worked there and he is still working there.)

      Past: After he *had worked* there for twenty years, they sold out. (Two events "completed"—twenty years' work and the sale.)

      Future: By the time she arrives, she *will have driven* 3,500 miles. (Part of the trip has been "completed"; part of it will "continue.")
      For additional discussion of the form of participles and tenses, see **Verb.**

2. Participles may be in either *active* or *passive* form:

Active: I *have chosen* Mary to do the work.
Passive: Mary *has been chosen* to do the work.

3. The present participle is often used as a noun: "*Swimming* is good exercise." "*Dancing* is fun." In such constructions, the participle is called a **gerund**, or verbal noun.
4. Both present and past participles are often used as modifiers (adjectives). For example, the *passing* parade, a *running* track, a *cherished* tradition, a car *purchased* on time payments.
5. When writers use participles in modifying phrases, they must be careful to avoid **dangling constructions**, i.e., allowing the phrases to modify words they cannot logically modify:

   Wrong: While *inspecting* the clutch, the throw-out bearing was found to be defective.
   Right: While *inspecting* the clutch, *I found* a defective throw-out bearing.

   Since the participial phrase has no subject, it can only modify the subject of the main clause, which in the first sentence is "throw-out bearing." See also **Dangling Constructions.**

PARTS OF SPEECH. See **Adjectives; Adverbs; Conjunctions; Interjections; Noun; Preposition; Pronoun; Verb.**

PASSIVE VOICE. See **Voice.**

P

PEOPLE, PERSONS. The word *people* usually connotes an anonymous group; *person* or *persons* refers to individuals thought of separately:

1. About 3,500 people attended the outdoor concert.
2. Five persons responded "yes" to the question.

PERCENT, PERCENTAGE. Although both words denote *per hundred, percent* should be used with a number only. When no number is expressed, *percentage* is used:

1. What percentage of the students are majoring in engineering?
2. About 38 percent of them are majoring in engineering.

Note that *percent* should be written out, not expressed as the symbol % (except in tables, illustrations, and mathematical expressions).

PERIOD. When used to denote a span of time, *period* is often overworked and used redundantly, e.g., *a period of two years*. Sometimes a writer may compound the felony by writing *a period of two years' duration.* The meaning *period* is clearly implied by the words *two years*. "He lived in New York for a period of two years" simply means, "He lived in New York for two years."

PERIODIC SENTENCE. See **Sentence Types and Construction.**

PERIPHRASIS. See **Inflated Expressions.**

PERSONS. See **People, Persons.**

PHOTOGRAPHS. See **Graphic Aids.**

PHRASES. See **Clauses and Phrases.**

PLAGIARISM. Plagiarism is the unauthorized use of the words or ideas of another. See *Avoiding Plagiarism* in **Library Research** (p. 163).

PLEONASM. *Pleonasm* is the use of unnecessary words. See **Inflated Expressions.**

PLURAL. Just a reminder never to use an apostrophe inadvertently in a simple plural noun: "I bought three *books* (not *book's*) yesterday."
The apostrophe is used, however, in the plural of individual letters and numbers:

1. There are four i's and four s's in Mississippi.
2. There are four 2's but only two 4's in eight.

See also **Apostrophe.**

POLY (as a prefix). See **Prefixes and Suffixes.**

POMPOUS LANGUAGE. Students, as well as technical, business, and governmental writers, should try to write as authoritatively as possible without using stiff or overly formal language. Not too long ago, formal writing—particularly business and governmental writing—fairly bristled with authority and stodginess. Business messages were sprinkled with such charming phraseology as *it is my hope, it has come to the attention of the undersigned, yours of the 5th inst.* In those days, no business or governmental writer would have dreamed of writing *I hope, I have learned,* or *your letter of March 5.*

Today, however, we have writers who *finalize* (instead of *finish*), *utilize* (instead of *use*), *initiate* (instead of *begin*). Sometimes, in an effort to write authoritatively, writers simply become stuffy. Just as OK may seem too informal in a business report, most of the following words may seem too full-blown:

Pompous	Natural
activate	start
assist	help
attenuate	reduce or lessen
commence	begin
endeavor	try
ephemeral	short-lived
implement	carry out or conduct
indigenous	native
magnitude	size or extent
mensuration	measurement
prior to	before
procure	buy or obtain
seek	look for
subsequent to	after or following
terminate	finish or complete

By itself, an occasional inflated word may do no harm, but too many of them will simply sound stuffy. Moreover, even ordinary words may appear ostentatious when a writer uses them in pompous constructions. Here is a simple statement inflated far beyond its worth:

> The work that transpired at the damsite prior to August of 1978 was presided over by Mr. John Jones, who retired on the thirty-first of that month. Subsequent to that date, responsibility was transferred into the capable hands of Mr. William Smith.

Simplified:

Until August 1978, John Jones was responsible for all work at the dam. On August 31, Mr. Jones retired and William Smith was put in charge.

Remember that standards of English usage are continually changing, and the trend in technical writing is toward a writing style that can be readily understood. Accordingly, for most exposition—technical, business, and governmental writing—writers should write with authority but without pretentiousness.

Compare with **Colloquial Language.** See also **Technical Terms and Jargon;** and *Avoid Highly Technical Words*, part 1, p. 15.

PORE, POUR. Simply remember, when it rains it *pours,* but you *pore* over a technical or other book. *Pore* is also a noun meaning the minute openings in the skin, a leaf, a rock, etc.

POSSESSIVE CASE. See **Apostrophe; Case; Pronoun.**

POST (as a prefix). See **Prefixes and Suffixes.**

PRACTICABLE, PRACTICAL. These two are often loosely used synonymously, but each has a distinct meaning. *Practicable* denotes a concept that has been neither tested nor proved but that appears feasible or possible. *Practical* refers not only to things but also to persons and implies usefulness:

1. A *practicable* idea for administering student registration (an idea that seems workable)
2. A *practical* person, system, or machine (as opposed to a dreamer or an outmoded system or machine)

PRE (as a prefix). See **Prefixes and Suffixes.**

PREDICATE. See **Subject and Predicate**.

PREDICATE ADJECTIVE, PREDICATE NOUN. See **Subject and Predicate.**

PREFIXES AND SUFFIXES. A *prefix* is a word or syllable placed before a word to create a word of another meaning. A *suffix* is a word or syllable added to the end of another word; here we are concerned with certain suffixes that change the meaning of the original word.

The most common prefixes are *all, anti, bi, co, cross, extra, half, infra, micro, mid, mini, mono, multi, non, over, poly, post, pre, re, self, semi, sub, super, supra, ultra, un, under,* and *uni*. Suffixes of concern include *elect, fold, like,* and *odd*. Other common words that can cause confusion are those formed with *great, in-law, quasi, vice,* and *well-known.*

The examples used in the following table are based on common word forms, and many exceptions to the general rules shown are not listed. Before using an unusual compound word, verify its form in an unabridged or college dictionary. See also **Hyphen.**

Note: Most prefixes used with (a) proper nouns or adjectives, (b) numbers, and (c) most two-word combinations are hyphenated. For example, *un-American, mid-European, pre-1930, post-1960, non-Spanish-speaking* (people), *post-Korean-War* (period).

1. Prefixes

Prefix	*Form*	*Examples*
all	compounds with all hyphenated	all-inclusive all-American
anti	solid except before words beginning with *i*	antienzyme antisocial anti-intellectual
bi	solid except before words beginning with *i*	biannual bicameral bi-iliac
co	solid except before words beginning with *w* (to prevent confusion with *cow*) or *o*, except cooperate and coordinate	coexist copolymer co-owner co-worker cooperate coordinate
cross	some combinations solid, others hyphenated, some two words; check usage in dictionary	solid: crossbeam crosscut crosstalk hyphenated: cross-country cross-examine cross-sectional (adj.) cross-section (verb)

P

Prefix	*Form*	*Examples*
cross *(cont.)*		two words: cross hair cross section (noun) cross vault
extra	solid when usage has given the compound a special meaning; hyphenated when used with other adjectives	solid: extraordinary extracellular extraterrestrial hyphenated: extra-hazardous (duty) extra-special (offer) extra-strong (solution)
half	often hyphenated, some solid, some two words; check usage in dictionary	hyphenated: half-hour half-life half-mast solid: halfhearted halftone (photo) halfway two words: half brother half note
infra	solid	infrared infrasonic
micro	solid	microcomputer microorganism microprocessor
mid	usually solid; two words with proper noun, but hyphenated with proper adjective	solid: midday midshipman midterm two words: mid Pacific hyphenated: mid-Pacific island
mini	solid	minicar minicomputer miniskirt
mono	solid	monoacid

Prefix	Form	Examples
mono *(cont.)*		monochromatic monovalent
multi	solid except before words beginning with *i*	multicolored multilateral multi-infectious
non	solid except with capitalized words	nonentity nonmetallic nonpartisan non-Germanic
over	solid when written as one word; hyphenated as a compound modifier	overachiever overestimate overrated over-the-counter (sale)
poly	solid	polychromatic polyconic polyvinyl
post	solid except with Latin words and proper nouns	postdated postgraduate posttest post-bellum post-mortem post–World War II
pre	solid except before capitalized words	prefabricated prefrontal preeminent preempt pre-Renaissance
re	solid except when hyphen is needed to prevent ambiguity	reenter recover (regain) re-cover (cover again) reform (improve behavior) re-form (form again)
self	hyphenated	self-evident self-locking self-serving
semi	solid except before words beginning with *i*	semiannual semimonthly semi-idle

Prefix	*Form*	*Examples*
sub	solid	subbase subcontractor subdivide
super	solid	superannuated supercargo superrefined superheated
supra	solid	supraliminal supramolecular supraorbital
ultra	solid except before words beginning with *a*	ultra-atomic ultramicro ultrasonic ultraviolet
un	solid except before capitalized words	unaffected unnerved un-American
under	solid when one word; hyphenated when joined with other words to create a compound modifier	underactive underrated under-the-counter (price) under-the-table (deal)
uni	solid	unicameral unicellular unipolar

2. Suffixes

Suffix	*Form*	*Examples*
elect	hyphenated except when name of office is two or more words	president-elect senator-elect vice-president elect
fold	solid except when used with numerals	tenfold threefold 30-fold

Suffix	*Form*	*Examples*
like	solid except when used with proper nouns, words ending in *ll,* and other compound words	catlike Einstein-like gill-like vacuum-tube-like
odd	hyphenated to prevent inadvertent humorous effect	thirty-odd singers 25-odd dollars fifty-odd books

3. Confusing Compound Forms

Word	*Form*	*Examples*
great	hyphenated	great-aunt great-grandfather great-great-grandmother
in-law	hyphenated	brother-in-law mother-in-law
quasi	two words when a noun; hyphenated when a compound adjective	quasi corporation quasi government quasi-scientific quasi-stellar
vice	usually hyphenated	vice-chancellor vice-president
well known	hyphenated when it precedes a noun	well-known person well-known theory She is well known.

P

PREPOSITION. A preposition indicates the relationship between a noun (the object of the preposition) and other words in the sentence. The preposition and its object make up a prepositional phrase, which can be either an *adverbial* or an *adjectival* modifier. The function of these phrases is identical with the functions of adverbs and adjectives.

Adverbial: They walked *along the stream.* (Phrase modifies the verb *walked.*)
Adjectival: He is a person *of high principles.* (Phrase modifies the noun *person.*)

Here are thirty of the most common prepositions:

about (the town)
above (the fence)
across (the street)
after (dark)
against (my principles)
along (the road)
around (the corner)
at (home)
before (breakfast)
behind (the house)
below (the surface)
beside (the river)
between (you and me)
down (the road)
during (the month)
for (the common good)
in (the laboratory)
into (the house)
like (a deer)
near (the river)
of (good character)
off (the track)
over (the barrier)
since (last year)
through (the years)
to (New York)
toward (the south)
under (the table)
with (your help)
within (six months)

PREPOSITIONAL PHRASES. Writers should be aware of two important facts about prepositions and prepositional phrases:

1. Prepositional phrases—particularly double prepositional phrases—are frequently overused. Writers often use long prepositional modifiers where simple adverbs or adjectives would suffice:

   Inflated: residents *of communities* located *along the coastline of Oregon* (three prepositional phrases in a row)
   More direct: residents of Oregon coastal communities

   In the preceding example, one word (*coastal*) neatly does the work of six words and eliminates two of the prepositional phrases.

   In the following example, three words eliminate the need for the inflated double prepositional phrase:

   Inflated: buildings covered *with ivy on the college campus*
   More direct: ivy-covered campus buildings

   Now, here is an example of two prepositions with a single object where no preposition is needed:

Inflated: a house made *out of adobe*
More direct: an adobe house

Although *out of* is commonly used in conversation, it is usually redundant. The meaning here is simply a house *of adobe,* or an adobe house.
Here is another common usage that is usually unnecessary:

Inflated: data *in regard to* the XYZ testing program
More direct: data *on* the XYZ testing program

Finally, note how a single adverb eliminates the need for a long prepositional modifier:

Inflated: He drove to the station *in a hurried manner.*
More direct: He drove *hurriedly* to the station.

See also **Inflated Expressions**; and *Watch for Redundancy and Circumlocution* in part 1, p. 17.

2. On the other hand, prepositional phrases may sometimes be necessary to clarify meanings. In some cases, writers may try to save too many words, substituting strings of noun modifiers for a prepositional phrase. Readers may then have to dissect the sentence to determine the meaning. For example:

Dense: The dye penetrant inspection results revealed no plate material surface defects.

A reader of that noun-packed sentence would probably have to analyze it and make up a prepositional phrase:

Clear: The dye-penetrant inspection revealed no surface defects *in the plate material.*
Dense: a 0–30 volt cathode ray tube power supply
Clear: a 0–30 volt power supply *for the cathode ray tube*

See also *Avoid Superconcise Writing* in part 1, p. 19.

PRESCRIBE, PROSCRIBE. To *prescribe* is to give directions, particularly medical directions in the form of a doctor's prescription. To *proscribe* is to forbid, prohibit, or outlaw.

PREVENTIVE, PREVENTATIVE. Simply take note that *preventive* is the simpler form and that *preventative* is really inflated.

PRINCIPAL, PRINCIPLE. *Principal* is often an adjective, although it may be a noun as well: *principal* parts, *principal* ingredients, *principal* and agent, school *principal,* interest on the *principal,* etc. *Principle,* however, is always a noun: *principles* of grammar, legal *principles,* etc.

You will never confuse them if you simply remember that the adjective, *principal,* is spelled with an *a.*

PROCESS EXPLANATION. The explanation of a process is usually arranged in chronological order, that is, as a series of successive events. A writer might have to explain a process to help readers determine whether it is practical, reliable, or more efficient than an alternative process.

Basically, there are three classes of processes: those performed by machine or a mechanism, those performed by persons, and natural processes, over which people have no control. Regardless of how the process is carried out, the explanation is written, usually in the third person, to describe the actions performed during the process. Sometimes a process performed by a person is described more informally—in the second person—which tends to personalize the discussion.

Process Performed by a Mechanism

In the following example, the writer explains what happens during the first stage of waste water treatment:

Primary Waste Water Treatment[12]

Primary, or the first stage of, waste water treatment removes some of the floating materials, substantial portions of the settleable solids, about half of the suspended material, and from one-quarter to one-third of the organic material requiring oxygen for decomposition and stabilization. Functions of the specific units involved in primary treatment are described in the following paragraphs.

The types and amounts of debris and miscellaneous materials that accompany sewage to a treatment plant include discarded shoes and clothing, tree limbs, and even small trees. Bar screens are used to intercept overly large debris from influent sewage. The debris is then removed to (1) prevent unsightly conditions in the receiving waters; (2) enable more effective disinfection, as overly large particles cannot be disinfected reliably; and (3) protect the mechanical devices used in the treatment process. Large pieces of organic

[12]State of California, Dept. of Water Resources, Bulletin 189, *Waste Water Reclamation: State of the Art* (Sacramento: Office of State Printing, 1973).

material are chopped up as they pass through the communitor, which is simply a grinder located in the waste water stream.

Grit chambers are used to settle and remove sand and other heavy or inert material, to protect pumps and other machinery, and to reduce the quantities of such materials that must be collected, handled, and stored during other phases of treatment. Pretreatment may also include aeration to clean organic material from the grit, to freshen the waste water, and to float oil, grease, and other solids so that they may be skimmed off.

After pretreatment, the waste water enters a sedimentation basin, where the rate of flow is reduced, and the quiescent conditions enable suspended materials to settle to the bottom. Plows or scrapers are used to move the settled material to a central point, where it is transferred to a digester for further treatment; meanwhile, the waste water, now minus the settled material, is disinfected in a chlorine-contact basin.

A digester is used to anaerobically (without oxygen) stabilize the solids removed from the sewage and permit their separation and concentration. The liquid portion is returned for reprocessing through the treatment plant. Periodically, the stabilized sludge is withdrawn from the digester and dewatered by permitting it to drain on sand beds, or by use of filters and centrifuges. Heat drying may also be used as the final step in solids handling.

Process Performed by a Person

When the process is carried out by a person, the explanation should not be confused with a set of instructions, which is written to enable another person to perform the process. Of the following two examples, the first explains the process, and the second provides step-by-step instructions for its performance.

Note the difference in the two forms: the writer explains the process in several paragraphs but presents the instructions as a series of numbered steps. Numbers are essential in instructions, because they enable an operator to determine exactly where each step ends and a new one begins.

Photographing Line Copy on the 6-A Process Camera

The 6-A process camera is a precision instrument designed to accurately reproduce original copy (called originals) for offset printing. To achieve the desired results, the camera operator must follow definite procedures as explained in the following paragraphs.

The first step is to determine the correct exposure time and required reduction, if any, for each item. This step is important, because it permits items receiving identical treatment to be grouped together (ganged) on the copyboard. Following this preliminary step, the camera operator will usually have several groups of originals, each of which will receive identical treatment.

After mounting the copy and moving the copyboard to shooting position, the camera operator is ready for focusing. To focus with the vernier scale in the camera bed, the operator sets the lensboard and copyboard indicators on their respective positions for the required reduction. Now the operator turns up the lights and swings the ground glass into place for the critical final focusing. The operator moves the lensboard forward or back, and up or down, as needed to center the image on the ground glass; this ensures the exact size of the image.

After obtaining the correct focus, the camera operator closes the shutter, turns off the lights, and swings the ground glass out of the way. Next, the operator attaches and centers the film on the back of the vacuum frame, applies the vacuum (which holds the copy rigidly on the frame), and raises the film carrier to exposure position.

The operator now sets the predetermined exposure time on the shutter control, and the copy is ready for photographing. On the Model 6-A camera, the operator depresses the shutter switch, which simultaneously turns on the lights and opens the shutter. The lights will remain on for the preset exposure time; when the lights go out, the exposed film is ready for developing.

Instructions

Here is the same material as step-by-step instructions:

1. Determine exposure time and reduction needed for each item.
2. Gang as many like items as possible on the copyboard and lock the frame.
3. Set lensboard and copy indicators to the required reduction.
4. Turn the lights up and swing the ground glass into place for final focusing.
5. Move the lensboard up or down and back or forth to center the image in the ground glass.
6. Close shutter, turn off lights, and swing the ground glass out of the way.
7. Center the film on back of the vacuum frame, turn on the vacuum, and raise the film carrier to exposure position.
8. Set predetermined exposure time on shutter control.
9. Depress shutter switch, which turns on lights and opens the shutter.
10. After the lights go out, remove exposed film for processing.

As you can see from the preceding instructions, there is a tendency to omit the articles (*the, a,* and *an,*), which may cause the writing to appear choppy. Although instructions should be concise, writers should not omit so many words that the instructions become unclear. For example, in step 2 (above), omitting *the* could result in a misreading. "Gang as many like items on copyboard and lock frame" could be misread as ". . . on *the* copyboard and on *the* lock frame." Whether to use articles in instructions is a matter of the writer's judgment.

One final word about instructions: they should be completely clear, particularly when equipment could be damaged or, more importantly, an operator could be injured if they are not clear. If special precautions are needed, writers should present them ahead of the step to which they pertain, not as an afterthought. Precautions should always be written in specific terms. Vague wording such as, "Do not apply excessive force when tightening the bolts," would leave readers wondering how much force is "excessive." Instead, the writer should provide specific information, including the reason for the precaution. For example:

Caution

Use a torque wrench to secure the head bolts. Do not exceed 30 ft-lb. A force greater than this will shear the bolts.

Natural Processes

As I mentioned in the introduction to this section, natural processes are not controlled by humans; rather, they respond to the dictates of nature. Accordingly, the best introduction to the explanation of a natural process might be a brief description of the scientific principles underlying the process. In the following explanation of color, the writer explains that the human brain, not the eye, deciphers the colors we see.

P

What is Color?[13]

Your eyes don't see the colors of a sunset—your brain does. Color is one of the brain's methods of organizing light energy to help you tell objects apart.

Light in the natural world is a vast mix of electromagnetic waves of different lengths. When light hits a surface, some frequencies are reflected off and some are absorbed. The frequencies that reach your eye from an object are the ones *not* absorbed, but reflected. If we see two objects, the one that reflects more long-wave light will appear redder; the object reflecting more short waves at us will seem bluer.

[13]From "What Is Color?" *Science Digest*, Summer 1980, p. 103. Reprinted by permission.

The light we can see is only a small sample of the rays that reach our eyes. Our brain divides that narrow span of visible frequencies into an internal code—the broad range of hues we call "all the colors of the rainbow."

How our visual system picks up and coordinates a jumble of energy pulses into green trees and red-breasted robins that keep their colors from one day to the next is a bizarre miracle only partly understood. Lining the inside of the eyeball opposite the iris is a layer of cells called rods and cones, which absorb light and send out nerve impulses. In each eye, there are 120 million rods—our black and white receptors—and 6 to 7 million cones, or color receptors. Cones come in three varieties: "red," "green," and "blue."

All three kinds of cones contain the same chemical, *retinal,* which serves as the eye's absorber of light. Oddly, in each type of cell, the same retinal absorbs a different frequency of light. In the red cones, it absorbs long waves; in the green cones, middle waves; and in the blue cones, short waves. Until recently, no one could say how retinal could act differently in each type of cell.

Dr. Barry Honig of the University of Illinois theorizes that the retinal molecule is surrounded by a different protein in each of the three types of cone, and that the arrangement of each protein's electrical charges makes the retinal molecule assume a different shape. The *shape* dictates which light frequencies it can absorb.

When sunlight strikes the cones of your eye, each retinal molecule "goes into a kind of bend," in Dr. Honig's words. This "bend" knocks the surrounding protein into a different shape, which in turn sets up five or more chemical reactions, which culminate in the whole cone cell sending a nerve impulse to the retina. There are over 100 million retinal molecules in one cone.

The retina's neurons "talk" to one another, comparing messages, and send information to the brain. The brain then decides, for example, that a fire engine is red. The process takes 0.01 of a second.

No one knows exactly how the brain makes its decision. "I'm amazed at how complicated it is," says Dr. Honig. "You could think of a simpler way to detect light. But nothing that's ever been devised by man is so sensitive under such a variety of circumstances."

PRONOUN. Pronouns are words used in place of nouns—*I, you, she, them,* etc.

1. Pronouns may be divided into eight classes:

   Personal: I, you, she, he, we, they, them, me, her, him, us
   Impersonal: it
   Reflexive: myself, yourself, herself, himself, itself, ourselves, yourselves, themselves
   Indefinite: each, either, neither, any, some, none, nobody
   Reciprocal: each other, one another

Demonstrative: this, that, these, those
Interrogative: who, which, what
Relative: that, which, who, whose

2. Pronouns can also function as adjectives; when so used, they are called *pronominal* adjectives:

   Which way did they go? (demonstrative)
   This book is interesting. (demonstrative)
   Some people dislike television. (indefinite)
   My dog is lost. (possessive)

3. *Case of Pronouns.* The *case* of a pronoun depends on its function. A pronoun may be (a) the subject (nominative case); (b) the object of a verb or preposition (objective, or accusative, case); (c) a possessive pronoun (possessive, or genitive, case). The form of a number of pronouns is determined by their cases:

   Nominative: I, he, she, it, we, you, they
   Objective: me, you, him, her, it, us, them
   Possessive: my, your(s), his, her(s), its, their(s), our(s)

   By heeding the case of a pronoun, you can avoid errors like these:

Incorrect	Case of Pronoun	Correct
John and *me* are going.	Nominative (subject)	John and *I* are going.
Me and *him* are going.	Nominative (subject)	*He* and *I* are going.
They invited Mary and *I*.	Objective (object)	They invited Mary and *me*.
Between *you* and *I*	Objective (object)	Between *you* and *me*
Us students will help.	Nominative (subject)	*We* students will help.
He asked *we* boys to help.	Objective (object)	He asked *us* boys to help.
All of *we* boys will help.	Objective (object)*	All of *us* boys will help.

P

*The subject is *all; us* is the object of the preposition *of.*

Them books are mine.	Nominative (modifies subject)	*Those* books are mine.

Note: With such combinations as "him and me," "you and I," "she and I," try the verb aloud with each pronoun alone, and you will "hear" the correct form. You would not say, "She told I" or "Me is going." The same is true of *we* and *us* when either is used with a noun—you would not say, "Us will help" or "He asked we."

See also **Who, Whom, Whoever, Whomever;** and **Who's, Whose.**

4. In technical and other expository writing, observe these conventions about pronouns:
   a. Use the nominative case after a linking verb when the pronoun has the same meaning as the subject. For example:

      The only student still in class was *I*.
      The only girls who received invitations were *they*.

      Hint: Begin the sentence (aloud) with the pronoun and verb and you will hear the correct form.
   b. Don't use reflexive pronouns (myself, himself, etc.) in ordinary subject or object forms. Use the reflexive form when (1) the subject and object are identical, or (2) as an intensive (for emphasis).

      Subject: James and I (not James and myself) will go.
      Object: He invited Mary and me (not Mary and myself).
      Reflexive: He (subject) berated himself (object) for not studying.
      Intensive: Helen, herself, handled all the arrangements.
   c. With conjunctions that connect such clauses as "Jim is heavier than I," determine the case of the pronoun by restoring (aloud) the words missing (but understood):

      Jane knows more about it than I (do).
      Jerry can run faster than I (can).
      She is as strong as if not stronger than he (is).
      I gave you the same grade as (I gave) him.
      He praised us more often than (he praised) them.

5. Make certain that a pronoun (a) agrees in number with its antecedent, and (b) refers to the correct antecedent. Those two problems are discussed in **Sentence Types and Construction.**

PROOFREADING, PROOFREADER'S MARKS. Proofreading is an important step that writers, particularly students, frequently ignore. For professional writers, proofreading is essential. If you are a student, it can earn you a better grade.

The best method for most writers is to have someone else read the paper. If that is impossible, try to leave at least a few hours between the time you finish the work and the time you turn it in. Somehow, it is almost impossible to whip a paper out of the typewriter and immediately find all the errors. A few hours later, on the other hand, you will find it easier to locate misspelled words, misplaced modifiers, incomplete sentences, illogical statements, and other errors. To spot spelling errors, read the paper "backward" (from right to left). To locate grammatical errors, awkward passages, and things you didn't mean to say, read slowly, sentence by sentence.

Here are a few of the more common proofreader's marks your instructor may use in correcting papers:

Symbol	Meaning	Example
ℯ/	delete	Don't worry about the cost.
∿	transpose	Teh price is too high.
⁐	close up	That is non essential.
#	leave a space	Thebook is lost.
¶	new paragraph	We landed in Germany. ¶ In France . . .
no ¶	continue paragraph	no ¶ We landed in Germany.
[	move left	[The book is lost.
]	move right	]The book is lost.
⊓	raise	The symbol for a footnote is *.
⊔	lower	The book is gone.
⊙	insert period	He lost his book I will lend him mine.
^,	insert comma	After they left the house was locked.
^;	insert semicolon	We're closed please come back tomorrow.
^:	insert colon	We need the following supplies pens, pencils, paper.
^-	insert hyphen	They bought first class tickets.
'	insert apostrophe	Johns book was stolen.

P

∨	insert quotation marks	She said, "I'm leaving for good."
sp	misspelled word	The work is divided into two catagories.
O	spell out	I need (2) books for that class.
/	lowercase letter	We're going to the Library.
≡	capitalize	We're going to the Fair Oaks library.
.... stet	let stand as it was originally	Don't worry about the cost ~~of the book.~~ stet

PROPER NOUN. See **Noun.**

PROPOSALS. At some point in your career, particularly if you work for a company that plans or constructs things, or both, you may become involved in a proposal. As the name implies, a proposal is a suggestion or request that some future action be taken. In most cases, a company or an agency will prepare a proposal as an offer to perform some task for another company, agency, or person.

More specifically, an aerospace company might propose an elaborate weapons system to the federal government, a proposal involving millions, or possibly billions, of dollars. At the other end of the scale, a business-communications firm might propose to install a word-processing system in a small company, involving a few thousand dollars at most.

For that matter, if you, as a homeowner, were looking for someone to paint your house, the offers you might receive from several competing paint contractors would constitute proposals, albeit minor ones, probably involving only a few hundred dollars. In another vein, if you, as an employee, were to suggest a new method of performing a task, you would be proposing a change in company operations.

Not long ago (in a western city), a group of bicyclists proposed that the city and county reserve bicycle lanes on a number of major streets. The movement grew into a large proposal to both the city and county governments. The result: an elaborate system of bike lanes throughout the city and county. As an offshoot of that proposal, another group of cyclists proposed that locked parking areas be provided in the central city for the growing number of people now cycling to work. That proposal resulted in a number of locked areas for bicycle parking.

All of the preceding offers and suggestions are examples of proposals of one type or another. As a technical writer, however, you will probably be more concerned with the first type mentioned: an offer by one company to perform some service for another company, i.e., a business proposal of some kind. For many companies, particularly large construction companies and defense contractors, proposals are a way of life. They depend on proposals for survival. A major proposal for such organizations may represent a joint effort by a team of specialists and can often cost thousands of dollars in preparation and production time.

Often, a company will receive a request for a proposal—known as an RFP—from the government or from another firm. In some cases, the RFP will specify a particular service or product. In other instances, the RFP may specify only a particular result, leaving the method or product up to the proposer or bidder.

Proposals vs. Reports

In one sense, a major proposal resembles a **report**. It must be organized efficiently, written competently, and presented effectively. It will usually contain many of the parts found in major reports: a cover, a summary, an introduction, tables and figures, and, often, a number of appendixes. On the other hand, there is a distinct difference between a report and a proposal: reports are most often chronicles of past accomplishments, tasks already completed. A proposal, however, is a suggestion for a future task, something the proposer hopes to carry out. If the proposal is to be successful, the proposing firm must convince the prospective customer that it can best perform the task or supply the desired service. Therefore, most proposals, as part of the proposed plan of action, contain information seldom found in reports—descriptions of the proposing organization's capabilities, facilities, special equipment, and specialized personnel. In addition, the proposal will contain a statement of the problem, a description of the technical plan offered to solve the problem, and, almost always, a schedule for accomplishing the work.

Parts of a Proposal

Not every proposal would be organized exactly like the one discussed here, but most major proposals will contain most of the parts described in the following paragraphs.

Summary or Covering Letter

The summary, the first contact with those who will evaluate the proposal, should be designed to capture their interest and orient them into the details that follow. It should include a brief statement of the problem, the approach proposed to solve the problem, and an estimate of probable success. The summary would contain few technical details and should be as brief as possible. In a recent proposal to the state of California (by an engineering firm proposing to undertake a needed study), the summary was a one-page covering letter signed by the firm's president.

Table of Contents

All proposals, except for the briefest sales proposals and brochures, should contain a contents page, enabling readers to find any part of the proposal quickly. The table of contents also provides evaluators with an overall view of the scope of the proposal. As with reports, the contents page is a directory to the entire package—it gives readers a quick look at the overall organization.

Here is a typical contents page. It was part of a recent proposal to perform a study of a coal-fired power plant, prepared in response to an RFP by a state agency. Note that the various headings serve as a directory to the key points in the proposal:

I. Introduction
II. Technical Discussion
  A. Statement of the Problem
  B. Coal Availability
  C. Coal Quality
  D. Coal Costs
  E. Action Plan for Supplying Coal
  F. Economic Evaluations
III. Tasks to be performed
  A. Phase I
  B. Phase II
IV. Project Organization
  A. The Project Team
  B. Project Control
  C. Project Facilities

V. Project Schedules
VI. Project Costs
VII. Qualifications
    A. General Corporate Capabilities
    B. Specific Experience and Capabilities
    C. Résumés of Key Personnel

Body of the Proposal

The body of a major proposal would contain, in one form or another, most of the information shown in the preceding table of contents. Here, in summary form, is how that information was presented in the proposal to perform a study of a coal-fired power plant.

Introduction. The introduction should demonstrate that the proposing organization fully understands the customer's requirements. In this particular proposal, the introduction stressed the benefits that would result from the study, pointing out that the state, which wants to construct several coal-fired power plants, must know:

a. whether the necessary coal supplies will be available for at least thirty-five years at a reasonable cost;
b. the best way to transport coal to the power plants;
c. how transportation costs will affect the cost of the electricity produced;
d. the environmental effects of a coal-fired power plant.

P

Technical Discussion. The technical discussion contained detailed analyses of each of the points mentioned in the introduction. The proposal promised complete studies of (a) the availability of coal; (b) the best type of coal, based on analyses of ash content, heat content, sulfur content, and toxic-element content; (c) the costs of coal; (d) the costs of transporting coal; (e) the costs of washing coal; and (f) environmental and public health concerns and costs. The proposal emphasized that the proposing firm would prepare detailed evaluations of those six points and develop a plan for supplying coal to the power plants.

Tasks to be Performed. This section of the proposal explained the detailed methods the proposing firm would follow to perform the tasks outlined in the technical discussion.

Project Organization. In this section, the proposing firm discussed the organization of the "team" it would establish to perform the study. The discussion included the specific positions the firm would set up, the controls it would devise, and the various levels of responsibility it would establish. This section also described the facilities the firm would use to perform the study.

Project Schedules. The schedules for the study included specific dates for meetings between the proposing firm and the prospective client, dates for completion of written progress reports, and a date for the final report at the end of the study. The proposal schedule was presented as a bar chart.

Project Costs. This section contained a detailed breakdown of specific costs for both phases of the study. The costs were based on present labor and overhead costs, with an estimated inflation factor included.*

Qualifications. This section presented specific details of similar studies performed by the proposing firm, along with detailed résumés of the members who would perform the proposed study.

Clarity in Proposals

As with reports, important considerations in proposals are clarity and correct use of language. Readers of reports may not be greatly disturbed by a badly written report; they may simply refuse to read it. What, however, might be the reaction to a poorly written proposal? Imagine for a moment that you were evaluating a proposal that was badly organized, carelessly written, and difficult to understand. A most probable reaction might be, "If they can't even discuss it intelligently, how do they propose to carry it out?"

Perhaps the old adage "Nobody is perfect" should not apply to proposals. Three recent proposals submitted to a state agency displayed evidence of the haste with which most proposals are prepared. In the covering letter of one, prepared on the company letterhead and signed by the company president, the word *proposal* was spelled *porposal*—unfortunately in the opening sentence. The reaction (by the recipient) to what

*Allowing for inflation can be most important in proposals, particularly if the work will extend over several years. Costs are almost certain to rise as time passes, a very crucial aspect of bidding.

might seem a trivial matter was, "Doesn't he even read what he is signing?"

The other two contained similar gaffes—although not on the first page the potential client would read—mainly errors in word usage. In one, *principle* was used as an adjective *(principal)* seven times. Although minor, errors like those may cause the prospective client to wonder about the competence of the proposing firm. A selling document should induce the client to accept the proposal, and if his or her attention is diverted by carelessness, he or she may have second thoughts.

A Check List for Proposals

Because proposals are most often prepared under tremendous pressure of time, certain parts can easily be neglected. A check list of things to look for would help:

1. Does the overall organization demonstrate a clear understanding of the potential client's problem?
2. Is the entire document unified? That is, does the organization hold the proposal together? This is particularly important in long proposals, where several people may have worked independently on various parts of the package.
3. Are all parts of the proposal written clearly? Will they be clear to the client? When time permits, check the proposal with one or two persons not too close to the actual document, and find out if they understand it.
4. Has the proposal been checked for small errors that may detract from its salability?
5. Does the proposal look like a selling document? What would be your reaction if it had been sent to you?
6. Are the cost figures reasonably accurate?
7. Is there a realistic time schedule for all tasks?
8. Will the proposal convince the prospective client that:
   a. You can solve the problem or problems?
   b. You can produce the desired result in the most effective manner?

PUNCTUATION. The principal purpose of punctuation is clarification—to indicate pauses (comma), stops (period and semicolon), questions (question mark), and introductory remarks (colon). Other punctuation marks indicate possession and contraction (apostrophe), exclamations (exclamation mark), compound words (hyphen), parenthetical separation (parentheses, brackets, and dashes), quoted material (quotation marks), and omissions (ellipsis).

You will find it easier to punctuate correctly if you understand the purposes and uses of the various punctuation marks. For discussions and examples of correct usage, see the following articles: **Apostrophe; Brackets; Colon; Comma; Dash; Ellipsis; Exclamation Mark; Hyphen; Parentheses; Question Mark; Quotation Marks; Semicolon.** Related articles include **Enumeration; Punctuation Marks, Spacing After;** and **Underlining.**

PUNCTUATION MARKS, SPACING AFTER. The conventions for spacing after punctuation marks are simple:

1. *Within a sentence,* space once after a punctuation mark, with these exceptions:
   a. Space twice after a colon.
   b. Do not space after:
      (1) a beginning parenthesis or bracket;
      (2) a period within an abbreviation.
2. *At the end of a sentence,* space twice after any punctuation mark (period, colon, question mark, or exclamation mark).

PURPOSELY, PURPOSEFULLY. To do something *purposely* is to do it intentionally ("on purpose"). To do something *purposefully* is to do it with a definite plan for reaching a goal, that is, determinedly.

PURPOSES. *Purposes* is often used as deadwood in such expressions as "I use this room for study purposes" or "I need this car for transportation purposes." In such uses, *purposes* is clearly redundant.

See also **Deadwood; Inflated Expressions;** and *Watch for Redundancy and Circumlocution*, part 1, p. 17.

Q

QUASI (as a prefix). See **Prefixes and Suffixes.**

QUESTION MARK. A question mark may be used in several ways:

1. To indicate a question:

How many books have you read?

2. To mark off a question within a sentence:

   Will they really lower taxes? was on everyone's mind.

3. To indicate (a) doubt about a date, number, or word; or (b) that a date or number cannot be established:

   a. The Mongols and their predecessors ruled North China for more than two hundred years (1140 through 1370? A.D.).
   b. For more than sixty years (? through 1278 B.C.), the Hittites battled the Egyptians.

4. A question mark is not used:
   a. following an indirect quotation:

      She asked whether we wanted the tickets.

   b. following a request written as a question:

      Will you please reply by February 10.

The use of quotation marks with question marks is discussed under **Quotation Marks.**

QUOTATION MARKS. To prevent confusion, quotation marks should be used correctly:

1. Use quotation marks to enclose every direct quotation; indirect quotations, however, do not require quotation marks:

   Direct: The manager said, "The test will be conducted tomorrow."
   Indirect: The manager said that the test would be conducted tomorrow.

2. Use single quotation marks to indicate a quotation within a quotation:

   The manager said, "The phrase 'initiate appropriate action' leaves a lot to the imagination."

3. When a direct quotation is more than one paragraph long, the quotation marks are placed at the beginning of each paragraph but at the end of only the final paragraph. When long quotations are indented on both sides of the page (from the left and right margins), the quotation marks may be omitted.
4. Use quotation marks to enclose the titles of subdivisions of published

works (parts, chapters, sections, etc.), and to enclose the titles of magazine articles, lectures, etc.:

a. He read the article "New Uses for Plastics."

b. The lecture "New Uses for Plastics" was interesting.

c. "New Uses for Plastics" is an interesting article.

Note: Titles of books, magazines, newspapers, and other published works are usually underlined (italicized). See **Documentation; Underlining.**

5. When a quotation mark and another punctuation mark appear together, apply these rules:
   a. Place a comma or period inside a quotation mark.
   b. Place a colon or semicolon outside the quotation mark unless it is part of the quoted material.
   c. Place any other mark inside the quotation mark when it is part of the quotation; place the mark ouside when it refers to a sentence of which the quotation is only a part. For example:

   Inside: Our motto is "Write for the Reader!"
   Outside: How do you like our motto, "Write for the Reader"?

R

RAISE, RISE. *Raise* is transitive and thus requires an object. *Rise* is almost always intransitive and takes no object. The principal parts of each are as follows:

Present	Past	Past Participle
raise	raised	raised
rise	rose	risen

You *raise* the required capital, but you *rise* each morning to go to work or to class.

RAZE. *Raze* is sometimes confused with *raise* but has an entirely different meaning. To *raze* is to tear down or demolish. When a building is razed, it is completely leveled, down to the foundation.

RE (as a prefix). See **Prefixes and Suffixes.**

RECURRENCE, REOCCURRENCE. A *recurrent* event or concept is one that occurs over and over again, as recurrent winter storms. In mathematics, a recurrent decimal is one in which a series of digits is repeated infinitely, as 0.333333 . . . etc. A *reoccurrence* is simply an event that occurs for the second time.

REDUNDANCY. See **Inflated Expressions.**

REFERENCE OF PRONOUNS. See **Sentence Types and Construction.**

REPAIRABLE, REPARABLE. Both of these adjectives refer to something that can be repaired or restored. *Repairable* usually denotes an object (motor, table, chair, etc.) that can be repaired. *Reparable* is correctly used to refer to a concept such as a person's reputation and is ordinarily used in a negative sense. For example: "The candidate's reputation was irreparably damaged by the accusations made against him."

REPETITION FOR EMPHASIS. Intentional repetition can be an effective device for emphasizing important information. There are several ways you can use it:

1. You can repeat a key word or words in a paragraph:

   An outline can be a writer's best friend. *It will* keep you on track as you write. *It will* remind you of everything you must include. *It will* give you a place to start and a place to stop.

2. You can set off sentences with numbers or letters and begin each with a key word or phrase:

   a. An outline will keep you on track as you write.
   b. An outline will remind you of everything you must include.
   c. An outline will give you a place to start and a place to stop.

3. You can provide a brief reminder, introducing it with some such phrase as *in brief, in short, to repeat.* In brief, use repetition to emphasize important information.

See also **Enumeration.**

REPORTS. Reports—in business and industry, or in government—are important to the success of many an organization. Business reports enable managers to keep track of company operations or to learn of new industrial developments. Reports are often the basis for crucial decisions, e.g., whether to increase production, develop a new product line, build new facilities. Research projects are always documented in reports. And every sales representative is familiar with sales reports.

Sometimes, reports may be the only product of a long, costly investigation. This is particularly true of governmental reports to legislatures. Ordinarily, governmental organizations neither construct nor sell anything. But they do investigate a wide variety of matters of public concern—air pollution, water quality (purity), pesticide use, mental health, drug use, population growth, and unemployment, to name only a few. All such investigations must be reported to the legislatures that authorized and funded them, and to the public.

Technical and other professional writers must be familiar with:

- the important qualities of reports
- informal and formal reports
- the parts of formal reports and how to prepare them
- the use of headings in reports

Those topics are the subject of this article. The article concludes with a brief sample report of the type students might expect to prepare. The use of tables, figures, and photographs in reports is discussed and illustrated under **Graphic Aids** (see p. 131).

What Makes Reports Effective?

To summarize in general terms, a report is effective insofar as it presents clearly the information readers need. Here are some of the important qualities of effective reports.

Three Outstanding Qualities

A report must be (1) accurate, (2) well organized, and (3) written in language that readers will understand the first time they pick it up. In those three respects, a writer's judgment becomes all-important. In part 1 (p. 5) I mentioned the author who suggests that expository writing is really an analytical task; that is, writers and authors must analyze their work to determine whether they are on-target to readers. In reports, which are usually intended for busy readers, this task becomes doubly important.

To analyze your writing, what are some of the questions you might ask?

Most of the following suggestions are obvious but are often overlooked by writers:

- Who am I writing for?
- How much do they know about the subject? Are they experts? Fairly knowledgeable? Novices?
- How much must I tell them if they are to understand the report?
- Does my report open with a clear introduction so that readers will understand the general subject immediately?
- Is the report clearly organized? Will the reader be able to follow it?
- Have I used language that readers will understand?

Other Important Qualities

The effectiveness of a report can also be measured by its thoroughness and objectivity.

Thoroughness. A report should cover all aspects of the subject and, to satisfy the needs of several types of readers, should be reasonably complete within itself. That is, it should contain enough information to enable all readers to understand it without reliance on outside sources. Some readers who are familiar with the subject may need only minimum information; others may be only vaguely aware of it; still others may know nothing at all about it.

A technical report on, say, water resources may be replete with such terminology as *annual primary recharge, artificial recharge, usable storage capacity, perched aquifer,* and *electrical conductivity,* to name only a few of the terms that might confuse even those with some knowledge of the subject. Here, a writer faces a dilemma. If he or she stops to define every term as it is used, the report may end up full of definitions. On the other hand, if he or she uses such terms, many of which have no simple definitions, without explaining them, the report will be useless to all but those with a first-hand knowledge of the subject.

One solution to that problem is to provide a glossary of terms as an appendix to the report (see *Appendixes*, p. 230). In the glossary the writer can define all technical terminology, using as many words as necessary, without slowing down every reader. Engineers who deal with such terms daily can pass up the definitions; a county supervisor—who might be an accountant—to whom the report is important can find the definitions in the appendix.

Another aspect of thoroughness is that a report should tell all sides of a story. Unlike advertising, for instance, which ordinarily presents only the positive side, a report must tell all, favorable and unfavorable. If there are unsolved problems, negative aspects, impending risks, costs that may escalate, etc., the writer cannot hide them or brush them off with a few summary statements. Writers may want to lead from strength—good news first—but they must always include the bad news as well.

For example, in a recent report discussing the possibility of desalting sea water, the writer presented a thorough discussion of how desalted sea water could be used to help alleviate the water shortage in the Southwest. The writer explained the techniques involved, discussed a number of successful small-scale experiments, and demonstrated that large quantities of ocean water could be desalted.

Then the writer presented the negative aspects: the high cost of the energy needed for distillation, the need to dispose of large quantities of salt that would accumulate (it cannot be returned to the ocean), the hot water that would return to the ocean (adversely affecting marine life), and the cost of distribution facilities to transport the fresh water to areas of need, most of which are many miles inland.

Objectivity. An objective report is based on concrete facts, with no intrusion of the writer's feelings. Regardless of the writer's enthusiasm—or lack of enthusiasm—for the subject, the report should simply present facts. Even the writer's conclusions (see p. 229), which are a part of some reports, should be based on objective facts presented in the report.

Objectivity requires that writers omit such adjectives as *tremendous, devastating, hopeless*. For instance, instead of writing, "The committee is hopelessly deadlocked," a writer should state objectively, "The committee has voted six times and still cannot agree on a decision." As another example, instead of discussing a "*tremendous* increase in the accident rate," an objective writer would present it as a "40 percent increase" (or whatever it actually was).

Headings in Reports

In most reports, headings serve as a writer's turn signals—they notify readers that the subject is shifting, with no need for awkward transitions.

When the subject changes abruptly, as often happens in reports, a writer may sense the need for some transition between thoughts. The use of a simple heading will usually bridge the gap neatly and concisely.

In a recent report on a new transistor, for example, the writer decided to introduce the device by first describing it and then explaining how it works. "Before discussing its function," he wrote, "it will be necessary to describe it briefly." Then, having described it, the writer compounded the felony by continuing, "Having described this new transistor, we now look at some aspects of its operation."

Two brief headings would have eliminated that awkward phraseology. All the writer needed was:

1. Description
2. Operation

Then, having finished the description, he could have used one heading to complete the transition to "operation."

Headings also help readers find information quickly, and they break up long pages of text.

Planning Headings

Headings are not difficult to plan. In fact, if you have developed an efficient outline, the outline topics may often double as headings in the report.

Levels of Headings

The positions of headings on the page indicate items of first-order importance, second-order, third-order, etc. Organizations issuing reports arbitrarily establish the positions they use to indicate the hierarchy, which can be seen very quickly in an outline. For instance, a *chapter heading* would correspond to a roman numeral (I, II, III, etc.), indicating first-order importance. The next level is shown by a *center* heading, corresponding to the capital letters in an outline (A, B, C, etc.).

The third order is indicated by a *side heading,* equivalent to the arabic numerals (1, 2, 3, etc.), which begins at the margin and is written in initial capitals and underlined. The fourth level (equivalent to a, b, c, etc.) is called an *indented heading;* it is indented, underlined, and followed by a period and the next line of text. If a fifth level is needed, it can be shown by a lowercase letter in front of an indented heading. Here is the hierarchy of headings down to the fourth level:

```
LEVEL 1           CHAPTER 3.  COOLING WATER SYSTEMS *

                 The purpose of the cooling water system in a
            steam-electric powerplant is to . . .

LEVEL 2                     Once-Through Cooling

                 In a once-through system, water is . . .

LEVEL 2                Closed-Cycle Evaporative Systems

                 Where water supplies are limited or . . .

LEVEL 3     Wet Cooling Towers

                 Two types of cooling towers, natural draft and
            mechanical draft . . .

LEVEL 4          Natural-Draft Towers.  A natural-draft tower
            is a large-diameter . . .

LEVEL 4          Mechanical-Draft Towers.  A mechanical-draft
            tower uses fans to provide . . .

LEVEL 3     Cooling Ponds

                 In a cooling pond, the heated water is . . .
```

Types of Reports

An exact list of every type of report would be difficult to compile, because most organizations issuing reports have differing classifications and names for them. In general, however, reports may be broadly divided into two classes: informal and formal.

Informal Reports

Informal reports are appropriate when the readers will be limited to those familiar with the subject, for example, a task force reporting to a manage-

*If a report is not divided into chapters, the first level can be shown by center headings in all caps; side headings indicate the second level, etc. (as in the sample report that begins on p. 233).

ment group, a laboratory worker reporting to a supervisor, a consultant reporting to a client. In such cases, the detailed information found in most formal reports would be unnecessary.

Informal reports include, among others, memorandum reports, letter reports, and blank-form reports.

Memorandum Reports. A memorandum report is precisely what the name implies—a report in the form of a memorandum. It is most often used for brief reports within an organization, where formality would be wasted. The format is similar to that of a memorandum; most are addressed to one or more persons from another person or from a group.

The organization of memorandum reports varies. Most are rather simple, containing up to ten or twelve pages. Occasionally, a memorandum report may be more elaborate; that is, it might be divided into sections, organized with several levels of headings (p. 219), and contain tables or figures, or both.

Here is the first page of a typical memorandum report:

M E M O R A N D U M

TO: Mr. Robert Jones
Mr. Paul R. Dean
FROM: Drainage Facilities Task Force

DATE: May 5, 1981
SUBJECT: Report on Acquisition of Rights-of-Way

I. ACTION TO DATE

The task force discussed these four questions on the acquisition of rights-of-way for the proposed drainage facilities in Merced and San Joaquin counties:

1. Can we establish the size of the drainage facilities now?
2. Can we determine the best location for them?
3. Can we set a firm date of need?
4. Can we establish a construction date now?

Our discussion of those questions ended in a qualified "no" to each. The discussion is summarized in Section III.

II. RECOMMENDATIONS

1. The task force cannot yet determine accurately the time of need for, or the size and location of, the required

R

Memorandum (cont.)

```
     facilities.  Therefore we recommend that the state
     should not acquire rights-of-way concurrently with the
     U.S. Water and Power Service for state drainage facili-
     ties.
 2.  We further recommend that an engineer from the Aqueduct
     Design Branch work with federal planners and designers.
     This will increase the likelihood that future state
     drainage facilities will be compatible with those being
     planned by the federal government.

               III.  DISCUSSION OF QUESTIONS

 (This section reported detailed discussions of the questions
 listed in Section I.)
```

Note that the sample memorandum report contained no detailed introduction but went directly to the heart of the matter. Since the recipients had established the task force, they were familiar with the subject of the report. The committee thus recognized that no detailed preamble was needed.

See also **Memorandums.**

Letter Reports. A letter report is an informal report in the form of a letter. It is often used for consultant reports to clients, for example. Because the recipient is almost always familiar with the subject, the writer needs no elaborate introduction, abstract, table of contents, or other parts of formal reports. Thus the letter report is another real time-saver. Ordinarily, a letter report consists mainly of words; most contain few tables, figures, or other **graphic aids.**

Blank-Form Reports. Blank-form reports are used within organizations to report progress of investigations, studies, laboratory experiments, etc. As implied by the name, the report is prepared on a preprinted form, often only one page. This type of report has one great virtue: if writers are confined to filling in the blanks, they will resist the temptation to ramble on at length. In fact, most writers welcome the opportunity to keep their reports brief and to the point.

A disadvantage of the preprinted form is its lack of flexibility. If unusual circumstances must be explained, a writer must often attach additional pages. In most cases, however, the blank-form report will fulfill its mission: to save time for both writers and readers.

The organization of such reports, particularly progress reports, is generally "past-present-future." This means the writer simply notes what has been accomplished, what is happening now, and what remains to be done.

Formal Reports

Formal reports include, among others, those issued by governmental agencies to legislators and to the public; technical reports to governmental agencies by contractors; and business reports to mangement, boards of directors, and stockholders. Most organizations specify what parts their reports are to contain. Although no universally agreed-on list of parts exists, a formal report:

- usually has a title page (discussed below), a preface (p. 224), and a table of contents (p. 225);
- is often preceded by an abstract (p. 226);
- usually has a brief summary (p. 227) and an introduction (p. 228) preceding the main body of the report;
- sometimes lists the author's conclusions (p. 229) or recommendations (p. 230) or both;
- may contain one or more appendixes (p. 230).

Let's look at those parts and some suggestions for preparing them.*

Title Page. The title page often causes problems for the writer, who must compose it, and the reader, who may not understand it. There is more to a good title page than merely a name and a date. The real problem with title pages, as you may have guessed, is the title itself: it is often too long, too short, or vague. An accurate, clear title is important not only to readers but also to librarians, who have to catalog the report.

The first thing to pare from the title is the deadwood. Ordinarily, you need no such wording as *a report on ________, a discussion of ________, a*

*The parts vary from organization to organization. Some may call the preface a *foreword*. In one report, the introduction may precede the first chapter; in another it might *be* the first chapter. Some reports might not have an abstract but may contain a summary as part of the introduction. Not every report would have a separate list of conclusions and recommendations, etc.

survey of ______. Since the document is obviously a report, the title need not repeat that fact. The word *introduction,* on the other hand, could be quite useful. Seeing the title, "Introduction to Scuba Diving," for instance, a reader would know instantly that the report contained basic information on scuba diving.

One-word titles are usually too vague. A fifty-page report titled "Rockets" or "Rocketry" would leave most prospective readers wondering, "What about rockets?" Since that subject would fill a fair-sized encyclopedia, it scarcely would do as a title for even a long report.

The ideal title provides some definite indication of the subject; for example, "A Copper Superconductor for Power Transmission" is both specific and indicative of the subject. If the title read, "A New Superconductor for Power Transmission," it would leave readers wondering about the nature of the conductor. "Water for Cooling Nuclear Power Plants," another specific title, should suggest that the subject will be the problem of obtaining water to cool modern power plants.

Often a two-part title will help remove the awkwardness from a long title. For instance, instead of "Wind Power as an Alternative Source of Energy," try: "Wind Power: An Alternative Source of Energy." That could also be an excellent two-line title, with no need for punctuation:

Wind Power
An Alternative Source of Energy

A two-line title will also help you eliminate double prepositional phrases, which are usually awkward. Instead of "Theory of Operation of Digital Computers," try:

Digital Computers
Theory of Operation

Although it may seem obvious, writers too often forget that an accurate report title is worth a bit of planning time.

Preface. Unlike a detective story, in which an author can beguile readers until the final page, a report should clarify its purpose on page one. And the preface is the ideal place to remove all mystery about the subject. Here you can tell readers—perhaps in two or three short paragraphs—what the report is about. You can prepare a preface by answering two simple questions:

1. What is important about the subject? (Why did you write the report?)
2. What does this report contain?

Some prefaces, particularly in government reports, contain the authorization for the report or acknowledgments of assistance from other organizations; in most cases, however, the answers to the two questions listed above will suffice. For example, a report on California soils for backyard gardeners had this brief preface:

THIS IS IMPORTANT	This report on California soils is intended for home gardeners, not for those who plant large acreages. Home gardeners should be aware that soils vary in different parts of the state and often in the same general area. Accordingly, the discussion is somewhat general and cannot provide the solution to every soil problem.
WHAT THIS REPORT CONTAINS	The topics discussed include soil problems, methods for improving soils, and fertilizing and watering techniques. In many communities, home gardeners can obtain more specific information from the local agricultural extension service.

In a longer report, the preface may be more elaborate but should contain much the same information. If a brief covering letter is substituted for the preface, it too should contain the type of information shown in the sample. (See the covering letter on page 233.)

One other point about prefaces is how writers begin them. A preface needs no wordy opening, such as, "The purpose of this report is to introduce readers to . . . etc." It should simply begin with a natural statement of intent: "This report will introduce readers to scuba diving."

Table of Contents. A complete table of contents is a specific directory to the report. Look at the table of contents to any textbook and note how quickly you can determine what the book is about. By reading the preface and the contents page, a reader should be able to learn the specific direction of the report.

The contents page should list headings to at least the third level (p. 220).

When the report contains figures and tables, list them separately following the last chapter. If the report contains appendixes (p. 230), list them after the tables and figures. Here is a sample contents page showing a common format. (The report from which it was taken was quite brief and not divided into chapters; instead, it contained five main sections.)

CONTENTS

LEVEL	I.	INTRODUCTION	1
I	II.	IMPROVEMENT OF SOIL STRUCTURE	3
		Organic Amendments	4
		Decomposition	5
	III.	SOIL PROBLEMS	7
LEVEL		Alkalinity	8
II		Acidity	8
		Shallow Soil (Hardpan)	9
	IV.	FERTILIZING	10
		Types of Fertilizers	11
LEVEL		Manures	12
III		Commercial Fertilizers	13
		Application	14
	V.	IRRIGATION	15
		Flooding and Ditching	16
		Sprinkling Systems	18

Tables

1.	Elements Required for Plant Growth	3
2.	Annual Application Rates for Fertilizers	15

Figures

1.	Irrigating by Ditching	17
2.	Typical Sprinkling System Layout	19

Abstract. An abstract is a digest of the report apart from and ahead of the introduction. For the convenience of librarians and others who file reports, many organizations include an abstract on a three-by-five card, which can be removed from the report and filed as a permanent record. By examining the abstract, readers can often decide whether a report will help them. The abstract, usually written in the language of the report, resembles a summary, except that it almost always is presented ahead of, or apart from, the report itself.

You should write the abstract after you have completed the report, not in advance. To prepare an abstract, find the most important *ideas* and *results* in the report. Then, think of the abstract as the report in miniature, and connect the ideas and results in as few words as possible, restricting yourself to 100 to 125 words.

Here is a typical abstract of a government report on cogeneration of heat and electrical power. The first paragraph presents the main ideas in the report. The second presents the results of the investigation reported.

ABSTRACT

Cogeneration is the simultaneous production of electricity and useful heat from the same fuel. The overall efficiency of a cogeneration plant is more than twice that of a conventional power plant. Cogeneration may well be one answer to the energy shortage; it can reduce the consumption of fossil fuels by one-third. Cogeneration will also reduce air pollution, thus contributing to a cleaner environment.

The Department of Energy has recently investigated the cogeneration potential of state facilities. Preliminary results indicate a potential of 1,500 million kilowatt-hours (1,500 gigawatt-hours) of electricity per year. Full development of this potential would save about 2 million barrels of oil per year.

Summary. The purpose of a summary to a formal report is similar to that of an abstract, i.e., to present the highlights without the details. Unlike an abstract, however, which is usually presented ahead of or apart from the report, the summary is part of it, often part of the introduction. Also, a summary is usually longer than an abstract; it often contains significant facts from each chapter or section. Thus, from the summary, readers can determine the full extent of the report before reading the details.

To prepare a summary, you must first determine the significant parts of the report. One way of doing this is to go through the entire text and underline the key points in each chapter. Then you can combine them into a brief summary. The sample introduction on page 228 contains a summary that appeared in a government report on weather modification.

Introduction. The introduction is the place to set the stage for things to come. In a concise introduction, the writer can provide the necessary background for the report, perhaps defining unusual terms, explaining special circumstances, establishing special provisions or limits. Although no exact formula for introductions exists, here is one arrangement that will help readers understand the report:

1. Background (if necessary)
2. Summary
3. Conclusions (if any)
4. Recommendations (if any)

That arrangement will help two types of readers acquaint themselves with the report: readers who are familiar with the background and those who are not. Thus, those who know something about the subject can read the summary, conclusions, and recommendations and move on to the following chapters, which contain the specific details. Those who are unfamiliar with the subject can bring themselves up to date by reading the background information.

The actual arrangement of an introduction depends on a writer's judgment and knowledge of the audience. Because reports are intended for many different audiences, a writer will have to analyze the circumstances surrounding the report and then decide what introductory information will be appropriate. Here is the introduction to the report on weather modification. Note how the writer used the final section of the summary to show what lay ahead in the four main sections of the report.

INTRODUCTION

There are two principal reasons behind attempts to modify the weather. One is to increase precipitation; the other is to reduce losses caused by fog, lightning, tornadoes, and hurricanes. Atmospheric experiments over the past thirty-odd years have been primarily efforts to increase or redistribute rain or snowfall. However, other experiments have been conducted to suppress hail and lightning, disperse fog, and modify hurricanes. All such experiments are classified as "weather modification."

Summary

Increased precipitation at the right time and place might mean the difference between success or failure of crops. Additional rain or snow might increase the output of a hydroelectric project, or enhance recreational opportunities by increasing the snowpack at a ski area. Increased rainfall might also aid the reproduction of fish by increasing streamflow when natural flows are low.

In another vein, adverse, and sometimes violent, weather may result in huge losses of property and losses of life as well. In the United States alone, annual losses from hurricanes average 25 million dollars. Moreover, hurricanes have killed some 15,000 Americans since 1900. Tornadoes also bring death and destruction each year; lightning and hailstorms cause large losses. Ground fog plays havoc with airline schedules and frequently causes serious highway accidents. If some means of controlling the weather were available, many such losses could be eliminated or at least reduced.

California has been interested in weather modification for some thirty years. Chapter 4, Division 1, of the California Water Code regulates the activities of the licensed contractors who conduct atmospheric experiments. A number of other states and the national government are also active in weather modification. This report describes a number of weather-modification projects conducted over the past thirty years—in California, in other states, and in other countries. The topics discussed include:

- A brief history of weather modification projects and the agencies conducting them (Section I).
- The theories behind weather modification (Section II).
- Various types of experiments, including precipitation increases, fog dispersal, hail and lightning suppression, forest-fire control, and hurricane dissipation (Section III).
- The negative aspects of weather modification: economic, ecological, and legal problems (Section IV).

R

Conclusions. This word has a special meaning in reports; it does not mean "the end." Rather, conclusions are the author's findings based on facts in the report. They are usually presented as brief summary statements in the introductory material, under a separate heading. For example:

Conclusions

1. Weather modification experiments cannot be conducted in a "vacuum." The production of rain or snow requires conditions favorable to precipitation, such as the presence of cumulus clouds.

2. Experiments to increase precipitation, suppress hail and lightning, and disperse fog have been only partly successful.
3. The only verified technique is that for dispersing cold fog. Experiments to disperse warm fog, which predominates in California, have been unsuccessful.

Recommendations. Recommendations are suggestions for action and, similar to conclusions, must be based on facts in the report. Ideally, they should evolve from the conclusions. Recommendations usually follow the conclusions under a separate heading. For example:

Recommendations

1. California should continue carefully controlled experiments to increase precipitation, suppress lightning-caused forest fires, break up hailstorms, and suppress fog.
2. Experiments to disperse warm fog at airports and along heavily traveled roads and freeways should be emphasized.
3. Experiments to increase rain or snowfall should be conducted only when there is no danger of flood.

Appendixes. Formal reports often include one or more appendixes, which contain supplementary or supporting information. On page 217 I mentioned the use of a glossary of terms. Writers may also want to present detailed mathematical tables, derivations of equations, compilations of data, and other information that might be cumbersome for many readers. If such information is presented in appendixes, those who want it or need it can read it; others can pass it up.

Appendixes are ususally lettered, e.g., Appendix A, Appendix B. Each appendix has a title and is preceded by a separate title page. If an appendix is quite long and divided with headings, it often has a separate table of contents. In the table of contents to the report, appendixes are listed by letter and title following the contents of the main report.

Planning a Formal Report

To plan a report, writers often use the brainstorming techniques discussed on page 11 in part 1. To satisfy the needs of potential readers, a writer could do worse than to anticipate their questions. For instance, if you were assigned to prepare a report recommenc 'ng that your company produce a new product, what questions might you expect? For now the order is unimportant, so let's brainstorm for a moment:

1. The new product
2. Cost of machines to produce it
3. Space needed
4. Skills of operators
5. Demand for the product
6. Has it been tested?
7. What are its advantages?
8. How does it work?
9. Is there a working prototype?
10. Will it be profitable?

Those are just a few of the questions you might expect to answer. Of course, the answers will have to be organized, but already you are working toward an outline.

A Specific Plan

To carry the planning one step further, assume that you must prepare a report on whether your company should manufacture a new automobile ignition system. Also assume that your department has been experimenting with it for several years and that your job is to convince the board of directors—most of whom are "financial types"—that the new system will be a profitable addition to your product line.

With the questions developed during brainstorming, you could try a tentative outline. If you, your colleagues, and your manager are convinced of the wisdom of the move, start from strength with this type of opener:

> For the reasons discussed in this report, the Production Division recommends that ABC Corporation manufacture and market this new system within one year.

Now try building an outline around that opener:

```
              NEW AUTOMOBILE IGNITION SYSTEM

SHOW THEM WHAT THE SYSTEM IS ALL ABOUT:
   I.  Introduction
       A.  Development of the new system
```

Outline (cont.)

```
        B.  Brief description
            1.  Principles of operation
            2.  Advantages over conventional system

CONVINCE THEM IT IS RELIABLE:
  II.   Reliability
        A.  Laboratory tests
        B.  Field tests

PROVE THAT IT WILL SELL:
 III.   Demand
        A.  Results of marketing research
        B.  Need to keep ahead of competition

DON'T HIDE THE PROBLEMS:
  IV.   Problems
        A.  New tooling
        B.  Skilled labor
        C.  Space required

DON'T FORGET THE COSTS:
   V.   Costs
        A.  Production
        B.  Marketing

NOW THE CLINCHER:
  VI.   Profit Potential
```

That outline illustrates only one suggested plan, and your colleagues may have other ideas for assembling the report. You might decide to put point VI (Profit Potential) first and develop the report around the profit projections obtained from the financial department. Writers often refine their plans several times before arriving at a final organization.

A Sample Report

The following report is an example of the type students might expect to prepare. It is brief and therefore not divided into chapters; rather, it consists of five main sections, each of which is indicated by a center heading (level 1). Note that the main sections, as well as the subsections, are identified in the table of contents.

The report illustrates (1) the use of a covering letter instead of a preface, (2) a table of contents, (3) the hierarchy of headings, and (4) the use of tables in reports. (The use of figures, photographs, and other graphic aids is discussed and illustrated under **Graphic Aids**.)

I thank R. S. Rauschkolb, University of California Soils Specialist, for use of the information in the sample report, which originally appeared in *Soil and Water Management for Home Gardeners,* Leaflet 2258, published by the University of California Division of Agricultural Sciences in 1976.

(To save space, the sample report was single spaced. In double space, standard for college work, the report would be ten pages long—about 2,500 words.)

John L. Jones
3742 Entrance Street
Fair Oaks, California 95628
May 15, 1980

Mr. Earl Bingham
American River College
4700 College Oak Drive
Sacramento, California 95628

Dear Mr. Bingham:

As you requested, I have enclosed a brief report on home gardening. The report is intended particularly for "backyard" gardeners, not for those who plant large acreages. Topics discussed include improving the soil, fertilizing, and irrigating.

Home gardeners should be aware that soils vary in different parts of California and often in the same general area. Accordingly, the discussion is necessarily general and cannot provide a solution for every problem that may arise. In many communities, home gardeners can get specific information from a reliable nursery and, often, from the local agricultural extension service.

Very truly yours,

John L. Jones

John L. Jones

Enclosure

R

CONTENTS

Tables

SOIL AND THE HOME GARDEN

INTRODUCTION

Of the many aspects of gardening, an understanding of the soil, along with the requirements for fertilizing and watering, is essential. The home gardener who knows how to manage the soil, when to fertilize, and how often to irrigate is commonly described as having a "green thumb."

The functions of a soil in relation to plant growth are to provide:

- support for plants,
- water and air for plant growth,
- a source of nutrients.

To obtain the best plant growth, the gardener must maintain or improve the soil by improving the soil structure, fertilizing when necessary, and irrigating on a regular schedule.

IMPROVEMENT OF SOIL STRUCTURE

Maintaining and improving the soil structure is one of the most important phases of soil management. It requires consistent and timely use of good management practices. For example, soil should be cultivated, or tilled, only when it has a medium moisture content, i.e., when it crumbles easily. If you stir the soil when it is too wet, it will puddle or pack. If you work it when it is too dry, it will form large clods or powdery dust.

Organic Amendments

Organic matter may be added to help maintain or improve the structure of most garden soils, particularly clay or adobe soils. Crop residues or grass clippings from a well-fertilized garden and lawn area provide a readily available source of organic matter.

Manures and composts are also valuable sources of organic matter. You can also use carbonaceous organic materials that decay slowly--peat moss, sawdust, rice hulls, shredded bark, or straw--either as a surface mulch or mixed into the surface soil. Since these materials are low in nitrogen, however, you should apply a chemical nitrogen fertilizer. Use about 1 pound of nitrogen (5 pounds of ammonium sulfate) for each

100 pounds of carbonaceous organic material used.

Decomposition

You cannot permanently build up large amounts of organic matter in the soil, because it soon decomposes and disappears. However, occasional additions of organic matter, either as plant residues or as manures, ensure a continuous supply of energy for soil organisms. As these organisms decompose the organic matter, they help maintain a satisfactory soil structure and also change the organic matter into inorganic nutrients that can be used by growing plants. It is through decomposition--the release of compounds that cement small soil particles together--that soil structure is improved.

FERTILIZING

Sixteen elements are required for plant growth and development. These are listed in Table 1.

Table 1. Elements Required for Plant Growth

Source		
Air and water	Soil	
carbon	nitrogen	magnesium
hydrogen	phosphorous	molybdenum
oxygen	potassium	manganese
	sulfur	copper
	zinc	calcium
	iron	chlorine
	boron	

As shown in Table 1, plants obtain the first three elements from air and water; they obtain the other 13 elements in inorganic form from the soil. Some plants--called legumes, e.g., peas and beans--are capable, with the aid of certain microorganisms, of obtaining the required nitrogen from the air. Other plants must obtain their nitrogen from the soil.

Soils are rarely fertile enough to supply all the nutrients required for optimum plant growth. On the other hand, soils are seldom deficient in all the required nutrients. California soils contain most of the elements essential to plant growth; therefore, only those lacking need be added.

Generally speaking, most gardeners tend to either under- or over-fertilize. Excessive fertilization is unnecessarily expensive and may result in damage to plants and undesirable changes in the plant environment.

Types of Fertilizers

Manure. When used correctly, manure can be an effective garden fertilizer. It helps improve soil structure and supplies a number of nutrients. However, manures may also contain weed seeds and a relatively high amount of salts.

Manures vary widely in nutrient content, depending on the type of manure and how it has been handled. Chicken manure is by far the most concentrated and, if used carefully, can serve as the only garden fertilizer. Dairy manure is far less concentrated and usually contains less nitrogen. On the other hand, steer manure from animals fattened on concentrated feeds is richer in nutrients than dairy manure. If it has been handled to prevent nitrogen loss, it may be the only source of nitrogen needed.

Commercial Fertilizers. Commercial fertilizers vary in the nutrients they contain. Ammonium nitrate, for example, contains only a single nutrient material. Ammonium phosphate, on the other hand, contains double nutrient compounds. Some commercial fertilizers are a mixture of nitrogen, phosphorous, and potassium. Federal law requires that the percentage of each nutrient be shown on the container.

Under this labeling method, the first number shown is the percentage of nitrogen (N); the second is the percentage of phosphorous (P), expressed as P_2O_5; the third is the percentage of potassium (K), expressed as K_2O. For example, 100 pounds of a 12-12-12 fertilizer contains 12 pounds each of N, P_2O_5, and K_2O. If a micronutrient has been added, it too, along with its percentage, must be listed on the container. Table 2 lists the nutrient content of a number of fertilizers.

Table 2. Nutrient Content of Fertilizers

Fertilizer	Percentage*		
	Nitrogen (N)	Phosphorous (P_2O_5)	Potassium (K_2O)
Chicken manure, dry	2.0-4.5	4.6-6.0	1.2-2.4
Steer manure, dry	1.0-2.5	0.9-1.6	2.4-3.6
Dairy manure, dry	0.6-2.1	0.7-1.1	2.4-3.6
Calcium nitrate (15.5-0-0)	15.5	0.0	0.0
Ammonium sulfate (21-0-0)	21.0	0.0	0.0
Ammonium nitrate (33-0-0)	33.0	0.0	0.0
12-12-12	12.0	12.0	12.0
Urea (46-0-0)	46.0	0.0	0.0

*P_2O_5 contains only 44 percent phosphorous, and K_2O contains only 83 percent potassium. The percentages of the oxide may be converted to percentages of the elements by multiplication. Thus, $P_2O_5 \times 0.44 = P$; $K_2O \times 0.83 = K$

Nutrient Deficiency

Nitrogen is naturally low in almost all California soils, and additional nitrogen is needed for optimum plant growth. In many lawns and gardens, it may be the only nutrient needed.

Phosphorous may also be lacking in some California soils, particularly in highly weathered soils, which are reddish and usually underlain with hardpan and claypan. Long-term cropping may also produce a deficiency of phosphorous.

Potassium is usually not needed, because most California soils naturally contain this element.

Sulfur deficiencies are uncommon but may exist in areas of heavy rainfall and where rainfall and irrigation water contain little or no sulfur. Some highly weathered soils may also be lacking in sulfur.

Zinc deficiency may be found in some areas, particularly where the surface soil has been removed during excavation and building. This can be corrected with a fertilizer containing zinc.

Iron deficiency is common when acid-loving plants are grown on soils containing lime (calcium carbonate). Iron deficiency can be corrected with acid or iron fertilizers, such as iron chelate or iron sulfate.

Application Rates

Suggested application rates for a number of common fertilizers are shown in Table 3. The rates given are general and will vary, depending on local soil conditions.

Table 3. Annual Application Rates for Fertilizers per 1,000 square feet

Fertilizer	Suggested Application Rate, lb
Chicken manure, dry	125
Steer manure, dry	450
Dairy manure, dry	600
Calcium nitrate (15.5-0-0)	16 to 25
Ammonium sulfate (21-0-0)	12 to 19
Ammonium nitrate (33-0-0)	7 to 12
12-12-12	20 to 35
Urea (46-0-0)	5 to 9

IRRIGATION

Water can be applied in furrows, basins, or with garden or lawn sprinklers; the latter are most often used by home gardeners. If sprinklers are used, the application rates should be low enough to permit the water to soak into the soil without running off. If it does run off, stop irrigating for an hour or so; then apply more water.

The texture and structure of the soil determine the amount of water it will hold at a single irrigation. The water-holding capacity of soil cannot be appreciably changed by small additions of organic material, e.g., peat moss or compost. However, large quantities of organic material mixed with soil can increase its water-holding capacity considerably.

The soil depth from which a plant normally extracts water depends on the rooting depth, which varies with the type of plant. In general, the main root zones of plants are as follows:

- lawn grass and leafy vegetables: top 1 foot;
- corn, tomatoes, and small shrubs: top 1 or 2 feet;
- small trees and large shrubs: top 2 or 3 feet.

Some home gardeners use plant symptoms as a guide to irrigation frequency. When short of water, many plants and grasses exhibit a dark blue-green color. Other plants may simply wilt; still others may do both. When those symptoms appear, it's time to water.

A general rule of thumb: irrigate thoroughly but not too frequently.

A FINAL WORD

In addition to the problems just discussed, the successful gardener must watch out for plant diseases, weeds, rodents, insects, and other pests. Information on plant disease control and pest control can be obtained from a reliable nursery.

Finally, the gardener with the "green thumb":

- fertilizes adequately but not excessively
- irrigates thoroughly but not too often
- promotes good soil structure by mixing in organic matter and by minimum tillage when the soil moisture content is medium wet.

RESTRICTIVE AND NONRESTRICTIVE MODIFIERS. *Restrictive* and *nonrestrictive* modifiers are not difficult to master, once you have recognized the difference between them. First, think of a *restrictive* modifier as a limiter—it limits, and thus "restricts," the meaning of the word(s) it modifies. Therefore, a restrictive modifier is not set off with commas; without the modifier, the statement would make no sense. First, we'll look at *relative* modifiers.

1. Relative modifiers are introduced by a relative pronoun—*that, which, who, whose.* For instance:

   Motorists *who drive recklessly* should be arrested.

   There, the limiting modifer *who drive recklessly* is essential to the meaning. Without it, the statement would mean, "Motorists should be arrested.".

   Now, let's rewrite the sentence with a nonrestrictive modifier:

   Reckless drivers, *who endanger the lives of others,* should be arrested.

   This time the limiting modifier is *reckless,* and the *who* clause merely adds additional information. Therefore, the clause is *nonrestrictive* (nonessential to the meaning) and is properly set off with commas. Remove it from the sentence and the meaning remains the same.

   Those two examples should help you identify the two types of modifiers. To repeat: the restrictive modifier limits the meaning, whereas the nonrestrictive modifier merely adds additional information. Here is another statement containing a restrictive modifier:

   Textbooks *that are out of date* should be destroyed.

   In a nonrestrictive construction, that statement would be nonsensical. "Textbooks, *which are out of date,* should be destroyed" might bring a cheer from many a student but would probably hit a snag in your local board of education.

   One last example will further clarify the difference between the two:

   Restrictive: The building *that once housed the Century Theater* will be restored. (Restrictive because the phrase identifies the building.)
   Nonrestrictive: The Mapes Building, *which once housed the Century Theater,* will be restored. (Nonrestrictive because the information about the theater is incidental; the limiting modifier is Mapes.)

   A question writers often ask is when to use *which* and when to use *that* in such constructions; I will also add *who*. When they are used as relative pronouns, there is no grammatical difference between *which* and *that*. As a general rule, however:

a. Use *that* with restrictive constructions: A boat *that ran aground yesterday* is still on the shore.
b. Use *which* with nonrestrictive constructions: John's boat, *which ran aground yesterday,* is still on the shore.
c. Regardless of the construction, use *who* or *whose* to refer to people; use *which* or *that* to refer to inanimate objects. You may also refer to people with *that;* never refer to people with *which*.

2. *Adverbial modifiers* may also be restrictive or nonrestrictive, especially those introduced with *where* or *when:*

Restrictive: Chicago is the city *where I enlisted in the Air Force.* (Restrictive because the clause is essential to the meaning.)
Nonrestrictive: I enlisted in Chicago, *where I had lived for twenty years.* (Nonrestrictive because the clause merely adds incidental information.)

Restrictive: Sales are heaviest *when everyone is looking for bargains.*
Nonrestrictive: Sales were heaviest right after Christmas, *when everyone was looking for bargains.*

RÉSUMÉ. See **Application Letters and Résumés.**

RHETORIC, RHETORICAL. *Rhetoric* refers to the theory and practice of composition, i.e., the structure and style of writing. *Rhetoric* is also used to mean propaganda or overelaborate and flowery speech. In this book, however, it has the first meaning—the structure and style of writing.

Rhetorical is an adjective meaning *concerned with style and effect.* The rhetorical mode of a piece of writing refers to how it is structured. A description usually follows a spatial pattern; an analysis is divided into distinct parts; the explanation of a process is usually in chronological order; etc. A number of rhetorical modes are discussed in the Handbook. See **Analysis; Classification; Comparison and Contrast; Definition; Description; Process Explanation.**

ROLE, ROLL. You play a *role* as an actor or in a particular function; in the army you attend a *roll* call. Your instructor may call the *roll* in class. Don't get caught—as did a governmental agency in a recent report—explaining the "roll of government."

RUN-ON SENTENCE. See **Sentence Types and Construction.**

S

SELF (as a prefix). See **Prefixes and Suffixes.**

SEMI (as a prefix). See **Prefixes and Suffixes.**

SEMICOLON. The semicolon marks a degree of separation between sentence elements greater than that shown by a **comma.** Use a semicolon:

1. To separate independent clauses that are not joined with a coordinating **conjunction.**
   a. *No connective expressed:*

      Ignition systems are assembled in Building A; radiators are processed in Building B.

   b. *Independent clauses connected with a conjunctive adverb:*

      Nobody is permitted on the ramp during the countdown; therefore, the safety check must be completed one hour before firing time.

      Note: The preceding clauses could also be independent sentences. Whether to create separate sentences or to use the construction shown is a matter of the writer's judgment. (See also **Adverb.**)

   Caution: Never separate two independent clauses with a **comma** alone or you will create a **comma splice.** Separate independent clauses with a semicolon or a period (and thus create a new sentence), or use a **conjunction** (*and, but, for,* etc.) along with the comma. See **Sentence Types and Construction.**
2. To separate sentence elements that contain smaller units separated by commas. See **Enumeration.**

Note: Do not use a semicolon to introduce a statement; the semicolon is a mark of *separation,* not of *introduction.* The latter is properly left for the **colon.**

SENTENCE FRAGMENT. See **Sentence Types and Construction.**

SENTENCE LENGTH. See *Control Your Sentences,* part 1, p. 21.

SENTENCE TYPES AND CONSTRUCTION. Fortunate is the writer who can construct effective sentences and connect them skillfully so that readers can follow the discussion without difficulty. Certainly the importance of clear sentences cannot be overemphasized.

I wish it were possible to provide every struggling writer with a magic formula for sentence construction. Someday, perhaps, some genius will invent a black box into which words can be inserted, a button pushed, and Presto! finished sentences in perfect syntax will emerge. Until that day, however, most writers will continue to struggle with words and word arrangement.

Fortunately, a number of guidelines for sentence construction are available, all of which should be helpful. First, let's examine four basic types of sentences: (1) simple, (2) compound, (3) complex, and (4) compound-complex. (Before continuing, you may want to review **Clauses and Phrases.**)

1. A *simple* sentence is merely a single independent clause: "Yesterday we signed a contract to build a bridge."
2. A *compound* sentence contains at least two independent clauses: "We have signed a contract to build a bridge, and we're going to hire a new construction crew."
3. A *complex* sentence contains one independent clause and at least one subordinate (dependent) clause: "Because we have signed a contract to build a bridge (subordinate), we're going to hire a new construction crew (independent)."
4. A *compound-complex* sentence contains at least two independent clauses and at least one subordinate clause: "We have signed a contract to build a new bridge (independent), but, before we can begin (subordinate), we'll have to hire a new construction crew (independent)."

Sentences may also be classified as *loose* or *periodic:*

1. A *loose* sentence opens with the main thought or thoughts—the subject—and may continue with subordinate ideas following the main idea. For example:

   A loose sentence (subject) follows the most common word pattern—subject followed by the verb followed by the object or complement.

   Here is a more complicated loose sentence:

   Waste-water treatment plants are frequently located at the lowest elevation in the collection system, adjoining a river, an estuary, or the ocean.

2. *Periodic* sentences lead up to the main idea, opening with a subordinate thought and following with the subject and verb at the end. Periodic sentences are particularly useful for emphasizing the link or connection between ideas. Here is the complete paragraph of which the preceding

example is only a part. The paragraph consists of four sentences; the first three are *loose* and the last is *periodic.* Note how the periodic opening connects the important thoughts in the last two sentences:

(Loose) Plans to reclaim waste water must include consideration of the treatment plant location and the need for distribution facilities. (Loose) For example, waste-water treatment plants are frequently located at the lowest elevation of the collection system, adjoining a river, an estuary, or the ocean. (Loose) Therefore, pumping plants and pipelines are often needed to transport the treated water to the point of reuse. (Periodic) When such additional facilities are required, the costs of construction, operation, and maintenance must also be considered.

Periodic sentences both lend variety to writing and connect ideas in adjoining sentences. Now, let's examine five important principles of sentence construction.

1. *Make certain that every sentence is grammatically complete, not a fragment or a run-on.*

Fragment: The customer complained that he had ordered nine books. And received only eight.

Analysis: The second sentence has no subject and should be part of the first.

Solution: The customer complained that he had ordered nine books but received only eight.

Run-on: The committee has arrived at several alternative solutions, it will report to the manager tomorrow.

Analysis: Two independent clauses joined with a comma (see **Comma Splice**).

Solutions:

a. Separate the clauses with a semicolon: The committee has arrived at several solutions; it will report to the manager tomorrow.
b. Use a period and begin a new sentence: The committee has arrived at several alternative solutions. It will report . . .
c. Connect the clauses with a coordinating conjunction: The committee has arrived at several alternative solutions, and it will report . . .
d. Delete the comma and use a compound predicate: The committee has arrived at several alternative solutions and will report to the manager tomorrow.

Incomplete sentences are often caused by carelessness. A sure preventive is a careful check of everything you write. See **Proofreading.**

2. *Make certain that you express parallel ideas in parallel form.* Parallel ideas are concepts with similar meanings or values. To express two similar ideas in similar form, you might use two nouns, two adjectives, two similar phrases, two similar clauses, etc. Parallel construction is convenient for readers because it helps them recognize like ideas and thus absorb ideas more quickly. Sentences with parallel ideas in parallel form are often called balanced. See also **Balanced Sentence.**

   The following examples contain ideas that should be in parallel form:

   Unbalanced: With a dimmer switch you may have bright lights for reading, lower intensity for dining, and then dim it even further for watching television.

   Analysis: Three similar ideas expressed in two noun phrases and an independent clause.

   Parallel: With a dimmer switch you may have bright lights for reading, lower intensity for dining, and even less intensity for watching television.

   Unbalanced: Masonite is the material on the front of the house, but the other three sides were constructed of stucco.

   Analysis: The first idea (Masonite) is the subject of the opening clause; the second idea (stucco) is in the predicate. Also, the writer mixed tenses (*is* and *was*) and voices—the first clause is active; the second is passive.

   Parallel: The front of the house is Masonite, whereas the other three sides are stucco. (two predicate nouns, present tense, active voice)

   Unbalanced: She said she would like to do something different, see new places, and she wants to meet new people.

   Analysis: Three similar ideas in two infinitive phrases and an independent clause.

   Parallel: She said she would like to do something different, to see new places, and to meet new people.

   Unbalanced: Apprentices must learn the use, maintenance, and how to repair refrigeration equipment.

   Analysis: Three similar ideas expressed in two nouns and an infinitive phrase.

Parallel: Apprentices must learn the use, maintenance, and repair of refrigeration equipment. (three nouns)

Unbalanced: The amount of loam may have to be experimented with before you can determine the correct amount to be used.

Analysis: The first clause is in passive voice, the second is in active voice, and *amount* is clumsily repeated.

Parallel: You may have to experiment to determine the correct amount of loam. (two infinitives in the active voice)

Unbalanced: In one day he not only selected a building site and then he obtained a construction permit.

Analysis: The correct idiom is *not only . . . but also*. The result here is two unbalanced clauses.

Parallel: In one day he not only selected a building site but also obtained a construction permit. (Two identical phrases following two correlative conjunctions; see also **Correlative Conjunctions.**)

3. *Make certain that the subject of a sentence or clause agrees (in number) with the verb.*
   a. *Simple subject:*

      Incorrect: The extent of the iron ore reserves are unknown.

      Analysis: The verb agrees with the object of a proposition.

      Revised: The extent of the iron ore reserves is unknown.

      Incorrect: Each of the boys attend a different school.

      Analysis: Each, the subject, is singular; the verb is plural.

      Revised: Each of the boys attends a different school. (OR) Both boys attend different schools.

   b. *Compound subject:*
      (1) When the two parts of a compound subject refer to the same person or thing, the verb is singular. For instance:

      In football, courage and stamina *is* a winning combination.

      The best writer and all-around student *is* David Mitchell.

      (2) When the second part of a multiple subject is connected to the

first by *along with, together with, as well as,* the construction is usually singular:

A heavy rain, along with high winds, is forecast for tonight.

The mayor, together with the city council, was on television last night.

(3) When two pronouns, or a noun and a pronoun, are connected with *or* or *nor,* the verb agrees with the word nearer the verb:

Either he or I *am* going.

Neither Jane nor we *are* going.

The solution to agreement problems: determine whether the subject is singular or plural, and make certain the verb is the same number.

4. *Make certain that a pronoun agrees (in number) with its antecedent (the word, phrase, or clause to which the pronoun refers).*

Incorrect: A trainee should be placed in the same work situation they will face on the job.

Analysis: Trainee, the subject and antecedent, is singular; *they,* the pronoun, is plural.

Revised: Trainees should be placed in the same work situations they will face on the job. (Note that the predicate noun [situations] must also be plural.)

Incorrect: The corporation failed to file their tax return on time.

Analysis: Corporation is a collective noun. Unless you are referring to individual members of a corporation, consider it singular. (See also **Noun.**)

Revised: The corporation failed to file its tax return on time.

Incorrect: She is one of those rare persons who is never late.

S

Analysis: The antecedent of *who* is *persons* (plural); therefore, both pronoun and verb are plural. (See also **One of Those . . . Who.**)

Revised: She is one of those rare persons who are never late.

5. *Make certain that a pronoun refers to the correct antecedent.*

Ambiguous: After the supervisor has noted all the alternatives available to solve the problems and analyzed them, she can choose the most effective one.

Problem: Does *them* (pronoun) refer to *alternatives* or to *problems*?

Solution: Get rid of the ambiguous pronoun.

Revised: After the supervisor has analyzed the problems and studied the alternative solutions, she can choose the most effective one.

Ambiguous: The manager discussed the order with Mr. Smith, and he said it might be a few days late.

Problem: Does *he* refer to the manager or to Mr. Smith?
Solution: Get rid of the ambiguous pronoun.

Revised: After discussing the order with Mr. Smith, the manager said it might be a few days late. (OR) The manager discussed the order with Mr. Smith, who said it might be a few days late.

Remember: Always use pronouns carefully. Readers have no time for guessing games.

SEXIST LANGUAGE. Students and all writers should be aware of a recent trend in language—an attempt to discuss men and women in equal terms and, where possible, to eliminate outright masculine references. Unfortunately, English has no singular pronoun that means *his and her,* and the use of *his* in situations where no gender is obvious has traditionally been acceptable.

For instance, "Each student must fill out his own application" is grammatically correct and has long been considered preferable to the frequently awkward use of the plural possessive *their*. The sentence can easily be changed to "All students must fill out their own applications." Now the sentence is grammatically correct, the meaning is identical, and the "sexist" pronoun has been deleted. In this case, the shift to the plural is more natural than the use of *his or her* with the singular noun: "Each student must fill out his or her application."

At times, avoiding use of a singular pronoun may result in awkward wording or lack of clarity. This is particularly true when the noun is clearly masculine, e.g., a boxer or a professional football player. In such cases, writers should use *he* and *his* when necessary. When they are simply citing examples, however, writers should try to alternate masculine and feminine pronouns. The use of plural forms, of course, usually eliminates the problem of gender reference.

Although some will scoff at such attempts to "desex" the language, writers should be aware that (1) standards of English usage are constantly changing and (2) many organizations are now encouraging, if not demanding, the elimination of sexist words. In many sentences, the removal of masculine implications requires only a bit of thought and a simple change in wording:

Original: A well contractor cannot guarantee that he will find water.
Revised: A well contractor cannot guarantee that water will be found.

Original: Make certain that the contractor and his crew are bonded.
Revised: Make certain that the contractor and the crew are bonded.

Original: Every person has a right to his own opinion.
Revised: All persons are entitled to their own opinions. (Note that in this case the predicate noun, *opinions,* must also be plural.)

Original: The average American prefers his steak medium-rare.
Revised: The average American prefers medium-rare steak.

Finally, writers can eliminate the suffix *man* from many words without awkwardness or loss of meaning. The purpose of the following suggested forms is to eliminate the connotation that certain roles are exclusively masculine:

Old Form	Suggested Form
businessman	executive, business representative
cameraman	camera operator
committee chairman	committee leader
committeeman	committee member
conference chairman	conference leader
department chairman	department head or chief
fireman	fire fighter
foreman	group supervisor or leader
insurance man	insurance agent or representative
man-hours	work-hours
manhood	adulthood
mankind	humanity or the human race
manned	staffed
pressman	press operator
salesman	sales representative

SHALL, WILL. *Shall* with the first person of verbs—I shall, we shall—is now considered obsolete by almost all authorities, and *will* may be used in all cases. About the only remaining uses for *shall* are to express:

1. *determination,* as in the sentence, "They shall not pass."
2. *commands and orders,* as in "The power shall be turned off before the wires are connected." This usage is common in military manuals, where *shall* is often used to indicate mandatory instructions.

SHOPTALK. See **Technical Terms and Jargon.**

SHOULD, WOULD. As with *shall,* the use of *should* (to mean *would*) in the first person is also fast becoming obsolete. *Should* is preferably used:

1. to mean *ought to.* In fact, the use of *should* to mean *would* might result in ambiguity: "I *should* be pleased to help" (when the writer means, "I *would* be pleased to help") might be interpreted as, "I *ought to* be pleased to help."
2. *to express uncertainty in a future event,* as in, "Should it rain tomorrow, the tournament will be postponed."

[SIC]. *Sic* is the Latin word for thus or so. It is often used within brackets to indicate an error in quoted material. For example:

The news story stated, ". . . the robbers crossed the Mexican boarder [sic]."

Here, [sic] was used to indicate the misspelled word *boarder.*

SITE. See **Cite, Sight, Site**.

SLANG. The vocabulary of slang comprises mainly nonstandard and usually short-lived words and expressions, most of which are unconventional (*endsville*), flippant (*buzz off*), irreverent (*fuzz*), metaphoric (*turkey*), and sometimes vulgar. Because slang words and phrases are usually intended to produce one or more of those effects, most soon become trite and begin to lose their edge.

On the other hand, a number of slang words and phrases have survived infancy and lived on to become part of the language, at least the spoken language. *Brass* (for high official) has been around for years, as has *grill* (question relentlessly). For some years now, thefts, questionable transactions, and out-and-out swindles have been termed *rip-offs.*

Many slang words are listed in standard dictionaries. A recent dictionary lists *rip-off* as slang; it lists *cool* meaning *excellent* as slang but designates *cool* meaning *composure* as informal. The same dictionary lists *brass* meaning *company official* as slang but designates *grill* (question relentlessly) as informal. A more recent expression that would seem to be slang is the *bottom line,* meaning the *result, culmination,* or *conclusion.* Yet, this same dictionary lists that definition for *bottom line,* without qualification as either slang or informal. Those few examples suggest the difficulty of distinguishing between slang and other informal language.

It would be difficult to select a slang word or phrase that would be appropriate in technical writing. In fact, unless used to create some very special effect, slang would be inappropriate in most expository writing.

Compare with **Colloquial Language.**

SO AS TO. Avoid this inflated phrase, which simply means *to.* For instance:

Inflated: Fasten the tiedowns snugly *so as to* prevent the load from shifting.
Concise: Fasten the tiedowns snugly *to* prevent the load from shifting.

SPELLING. Careful writers spell words correctly so that readers are neither deceived nor compelled to determine exactly what the writer is trying to say. If, for example, you write *quite* when you mean *quiet, assistants* when you mean *assistance,* or *personal* when you mean *personnel,* readers may give up on you. In addition, too many misspelled words may cause readers to view a piece of writing with a natural question: "If this writer can't even spell, does he know what he's talking about?"

The main reasons for the difficulty we all experience in spelling English words are that (1) many of our words do not sound the way they are spelled; (2) many of the same sounds are spelled differently; and (3) many common words are mispronounced, and the mispronunciation creeps into the written form.

The first problem stems from the fact that although printers began to standardize English spelling some 300 years ago, nobody has ever completely standardized it to harmonize with the spoken word. For instance, each of the following words ends in *ough,* yet each is pronounced differently: *bough, cough, though, through.* (Try them out loud and note the four different sounds for *ough.*)

The second problem—similar sounds with different spelling—is probably the cause of most of our spelling woes and agonies. First of all, we have varying suffixes:

1. *able* and *ible:* advisable and changeable BUT flexible and credible.
2. *ant* and *ent:* tolerant and resistant BUT confident and independent.
3. *ance* and *ence:* attendance and complaisance BUT independence and existence.

If you try all twelve of those words aloud, you will hear that, despite the variation in spelling, the suffixes in each group sound the same.

Then, there are the many **homonyms** in the English language—the words with different meanings that are pronounced alike. How often we run into these:

affect and effect

capital and capitol

complement and compliment

council, counsel, and consul

discreet and discrete

pair, pare, and pear

plain and plane

principal and principle

stationary and stationery

there, their, they're

Question: Can you distinguish between each of those pairs and trios of homonyms?

The problem caused by homonyms is pointedly illustrated by the following sentences:

1. While looking for a building *site* away from the *sights* of the city, Mr. Brown was *cited* for trespassing.
2. The new *edition* of the dictionary will be a valuable *addition* to your library.
3. The group was *all ready* to go but the bus had *already* departed.
4. She *alluded* to one of Newton's laws but her meaning *eluded* me.
5. His *ascent* to the presidency met with general *assent*.

The third category—words commonly mispronounced—includes such ordinary terms as accidentally (not *accidently*); athletics (not *atheletics*); height (not *heighth*); irrelevant (not *irrevelant*); mathematics (not *mathmatics*); quantity (not *quanity*); temperature (not *temperture*); nuclear (not *nucular*).

Occasionally, common mispronunciations can lead to the use of nonexistent words. Not long ago, I read a newspaper article in which the term *renumeration* [sic] appeared. The story, a staff-written article on real-estate regulations, was quite detailed and apparently based on considerable research. Unfortunately, the main point I remembered was the use of *renumeration* (no such word exists) instead of the intended word, *remuneration*. The moral: don't use terms that can bring smug smiles to readers' faces.

Are you already wondering what you can do about this problem? Fortunately, the solution is fairly simple—if you are willing to invest about 50 cents and a bit of effort.

Although there are rules and formulas—*i* before *e*, doubling consonants, etc.—for English spelling, there are also exceptions to all those rules, and few of us could actually memorize all the rules and exceptions. A more practical solution for solving spelling problems: never take a chance! Whenever you have the slightest doubt about how a word is spelled, take an extra minute and look it up in a good dictionary.

Next, invest about 50 cents in a small pocket notebook, and use it systematically to capture all the words that fool you. Every time you miss one, or are forced to look one up, jot it in the notebook. Then, the next time you are undecided about a word already in your notebook, look in the notebook. Before long, most of those troublesome words will be "yours," and you will soon discover that spelling is not the chore you once thought it to be.

However, the time to attend to spelling errors is while you are revising a first draft. If you stop to look up too many words as you are writing, you will break your concentration and almost certainly lose some of your spontaneity.

See also **Proofreading**; *Revise, Revise, Revise* in part 1, p. 32.

SPLIT INFINITIVE. A *split infinitive* is one that is interrupted by an adverb. Although its usage was long considered a high crime—or at best a misdemeanor—a split infinitive is now recognized as the natural form for many expressions. Because an adverb modifies the verb, its natural position is next to the verb. For instance, to change the position of the adverbs in the following sentence would result in awkwardness:

The purpose of the special classroom drills is to gradually increase word comprehension and to eventually eliminate most spelling deficiencies.

If you try that sentence out loud, you will "hear" that the word order is quite natural (rather than ". . . to increase comprehension gradually and to eliminate most spelling deficiencies eventually.")

On the other hand, writers should not split infinitives with phrases or clauses. For instance:

Awkward: I eventually learned to, although it took several months of practice, spell a great many words.
Improved: Although it took several months of practice, I eventually learned to spell a great many words.

SQUINTING MODIFIER. See *Be Careful with Modifiers*, part 1, p. 20.

STATIONARY, STATIONERY. The first of this pair is the adjective *stationary* (meaning fixed in place). Simply associate *stationery* with paper, and the *er* in both words will help you remember the correct spelling for the noun (*stationery*) and, by elimination, the adjective (*stationary*).

SUB (as a prefix). See **Prefixes and Suffixes.**

SUBJECT AND PREDICATE.

1. Subject
   a. In its simplest form, the subject of a sentence or a clause is a noun or pronoun:
      (1) *Fish* abound in this stream.
      (2) *I* am coming home tonight. *His* is the better automobile.
      (3) (subordinate clause) *who* has been nominated for the senate.
   b. However, the subject may take other forms as well:

      Gerund: Seeing is believing.

      Prepositional infinitive: To know her is to love her.

      Bare infinitive: Let them go.

      Adjective used as a noun: The wealthy pay the highest taxes.

      Other parts of speech: Hardly is an adverb.

      Group of words: Six plus two is eight.

      Clause: Whoever did that should be punished.

      Subject understood: Thank you (meaning *I* thank you). Come in (meaning *you* come in).

   c. Compound subject. A compound subject comprises two or more nouns, noun equivalents, or pronouns. The verb following a compound subject is usually plural. However:

(1) When the two parts of a compound subject refer to the same person or thing, the verb is singular.

(2) When the second part of a multiple subject is connected to the first by *along with, together with, as well as,* etc., the construction is often singular.

(3) When two pronouns, or a noun and a pronoun, are connected with *or* or *nor,* the verb agrees with the word nearer the verb.

For specific examples illustrating those three rules, see **Sentence Types and Construction.**

2. The predicate is that part of a sentence or clause that tells what happens to the subject; it consists of a verb with its complements and modifiers. The predicates in the following three examples are in italic type:

a. I *am coming home tonight.*

b. Fish *abound in this stream.*

c. (subordinate clause) who *has been nominated for senator.*

A noun or adjective that completes the meaning of a linking verb is called a predicate nominative or a predicate adjective, respectively.

predicate nominative: This tool is a *spanner wrench.*

predicate adjective: This tool is *essential.*

See also **Compound Predicate; Sentence Types and Construction.**

In a complex sentence, both the main clause and the subordinate clause have subjects and predicates. In the following sentence, the subject and predicate of the main clause are italicized; those of the subordinate clause are italicized and underlined:

Mr. Brown, <u>*who has been nominated for senator,*</u> *will speak tonight.*

S

SUBJECT AND VERB AGREEMENT. See **Sentence Types and Construction.**

SUBJUNCTIVE MOOD. See **Mood.**

SUBORDINATE CLAUSE. See **Clauses and Phrases.**

SUBORDINATING CONJUNCTION. See **Conjunction.**

SUPER (as a prefix). See **Prefixes and Suffixes.**

SUPERCONCISE WRITING. See *Avoid Superconcise Writing*, part 1, p. 19.

SUPRA (as a prefix). See **Prefixes and Suffixes.**

SYMBOLS. See **Abbreviations.**

SYNTAX. This rather impressive-sounding word simply refers to the arrangement of words in phrases and sentences. Particularly pertinent are the discussions of conciseness, modifiers, and sentences in part 1 (pp. 17–25) and **Sentence Types and Construction.**

T

TECHNICAL TERMS AND JARGON. Consider the following sentence written by a demographer, a person who specializes in population studies. In this instance, he was writing about the effect of a sudden population increase in Oroville, California, which became a boom town in the 1960s, when the state of California was constructing a large dam on the Feather River.

The demographic makeup of the populace is important from the standpoint of service requirements, such as the need for schools, hospitals, etc.

Perhaps this sentence would be meaningful to another demographer. For those unfamiliar with the subject, however, the sentence would be indirect (*demographic makeup*), vague (*service requirements*), inflated (*from the standpoint of*), and slightly pompous (*populace*). Even worse, the sentence contains only a suggestion of the real meaning the writer intended.

Now let's write what the writer should have written:

The size, growth, density, and age groupings of the population help determine the need for schools, hospitals, recreation facilities, public transportation, and police and fire protection.

Now we know what the *demographic makeup* (size, growth, density, and age groupings) and the *service requirements* (schools, hospitals, recreation facilities, public transportation, and police and fire protection) actually are.

Why was the original sentence unclear? Because most of us are unacquainted with *demography* and what it means. In this case, the writer forgot that many readers would not have a demographer's knowledge of the subject matter.

Sometimes, of course, the use of a technical term may be advisable. Often, a single technical term may put across an idea that might otherwise require a long, tedious explanation. Take, for example, the term *miter,* which means to bevel, usually at 45 degrees, each of two surfaces to be joined to form a 90-degree corner. It is with words like this that writers must exercise considerable judgment.

If you were the writer, could you simply write *miter the corners* and assume that all readers would know what you mean? Or would a detailed explanation and perhaps an illustration or two be better? Obviously, such a question cannot be answered with a simple yes or no. Here is a case where you would have to consider who your readers are and decide for yourself how much you must tell them.

The same principle can be applied to *jargon,* or *shoptalk,* which is language peculiar to a particular trade or profession. The language of printers, for example, is replete with such colorful terminology as *picas, ems, ens, goldenrod, mutts, quads,* and *thins.*

A good rule of thumb: avoid shoptalk and highly technical language unless (a) you know that all your readers will understand it, or (b) you cannot express a particular concept in ordinary terms. And if you must use technical words, be sure to explain them, unless you know that all of your readers will understand them.

See also *Avoid Highly Technical Words*, part 1, p. 15.

TENSE. See **Verb.**

THAT. *That* may be an adjective, an adverb, a conjunction, or a pronoun.

Adjective

1. I want *that* book.
2. *That* house is for sale.

Adverb

1. Are you really *that* desperate?
2. He is actually *that* busy.

Conjunction

1. He worked so long *that* he actually collapsed.
2. I told her *that* I would be there.

Pronoun

1. *Demonstrative*
   a. I'll be there. *That* you may be sure of.
   b. He often smacks his lips. *That* is annoying.
2. *Relative*
   a. She lives in a house *that* faces the river.
   b. There are three things *that* I must do.

Note: That as a conjunction or relative pronoun may often be omitted with no loss of meaning:

1. I told her (that) I would be there.
2. There are three things (that) I must do today.

The trend today is to omit *that* in such constructions and to simplify weighty relative modifiers:

1. water *that is high in* mineral content = *highly mineralized* water
2. a house *that faces the* river = a house *facing* the river
3. a practice *that is* well known = a *well-known* practice

For the use of *that* in nonrestrictive modifiers, see **Restrictive and Nonrestrictive Modifiers.**

THEIR, THERE, THEY'RE. *Their* is the possessive form of the pronoun *they; there* is an adverb meaning *in that place; they're* is the contraction for *they are:*

1. Writers should select *their* (possessive) words carefully.
2. He is waiting over *there* (adverb).
3. *They're* (they are) all taking an exam today.

THERE IS, THERE ARE. See **Expletive.**

THESIS. The thesis of a report, theme, or other expository paper is the central statement of what the writing is about. It should cover the entire subject briefly, telling the reader what the general area of coverage will be. It may be a single sentence, or it may be two, three, or four sentences. The thesis should state the central purpose of the text and hold together the entire presentation. It is literally the backbone of the paper.

For example, in this paper describing the chemical structure of water, the writer began with the thesis that water is no longer considered a simple chemical compound, H_2O:

For many years, water was considered a simple chemical compound whose formula was represented by H_2O, or two parts hydrogen and one part oxygen. Today, however, that concept of a water molecule has changed, because two new types of hydrogen have been found in water. One of these, called deuterium, is twice the weight of the normal hydrogen atom and combines with water in a 2–1 ratio to form D_2O. The other, called tritium, is three times the weight of ordinary hydrogen and combines with oxygen to form T_2O.

Now, having advanced that thesis, or central idea of what the theme is about, the writer can continue with the significance of the new discoveries.

Here is a thesis introducing a report on home gardening:

Of the many aspects of gardening, an understanding of the soil, together with the requirements for fertilizing and watering, is essential. The home gardener who knows how to manage the soil, when to fertilize, and how often to irrigate is often described as having a "green thumb."

With that statement of "aboutness," the reader should now expect to find full information on the three items named in the thesis: soil management, fertilization, and irrigation.

TO, TOO, TWO. Writers invariably have trouble with *to* (preposition and sometimes an adverb), *too* (the adverb meaning excessively, also, very), and *two* (the number). I cannot recall a substitution of *to* or *too* for *two,* but I have often read such sentences as "Ten boys are *to* many *too* control." In that sentence, *to* is in *too*'s place and vice versa.

T

TOPIC SENTENCE. The topic sentence of a paragraph is a general statement of the paragraph's controlling idea or central thought. In one sense, it is a miniature thesis for the paragraph (see **Thesis**). The sentences that follow it should directly support the topic statement. See also *Create Meaningful Paragraphs*, part 1, pp. 25–32.

TRANSITIVE AND INTRANSITIVE VERBS. See **Verb.**

U

ULTRA (as a prefix). See **Prefixes and Suffixes.**

UN (as a prefix). See **Prefixes and Suffixes.**

UNDER (as a prefix). See **Prefixes and Suffixes.**

UNDERLINING. Underlining, which corresponds to italicizing in printed copy, is used to indicate (1) titles of books, magazines, newspapers, and other printed material; (2) foreign words that have not been anglicized; (3) special emphasis; and (4) words to be considered not for meaning but as words. Some examples:

1. *Titles.*

   Book: General Chemistry
   Magazine: Time
   Newspaper: Washington Post (Note that the city is part of the title.)
   Report: Measuring Solar Radiation

   Note: The titles of articles in magazines and newspapers, stories in anthologies, etc., are set off with quotation marks. (See **Quotation Marks.**) For a complete discussion and examples of the use of underlining and quotation marks in footnotes, references, and bibliographies, see **Documentation.**
2. *Foreign words.* Foreign words that have not been anglicized are often underlined (italicized when printed). Words and phrases such as nouveau riche, sine qua non, tête-à-tête are usually underlined. Other common words of foreign origin, such as éclair, buffet, naive, chic, are almost always written with no underlining. See **Foreign Words.**
3. *Special emphasis.* Words are sometimes underlined to indicate special emphasis. For instance:

   a. The mixture must be stirred for five minutes.
   b. That was some experiment.

   Although the practice is quite common in informal writing, too many underlined words may actually detract from the emphasis. A better method might be to use a locution such as "Note carefully" or "Note especially" or to indicate the importance of the items by setting them off in special constructions. For examples of this last type of emphasis, see **Enumeration; Repetition for Emphasis.**

4. *Words to be read as words.* Words are sometimes underlined so that readers will consider them not for meaning but as words. For instance:

   a. Percent should be used with a number only; when no number is expressed, use percentage.

   b. Like means similar to or desirous of.

   See also **Italic Type, Italicizing.**

UNI (as a prefix). See **Prefixes and Suffixes.**

UNIQUE. Strictly speaking, *unique*—derived from the Latin *unicus* (meaning *one*)—denotes *the only one of its kind.* Although modern usage has broadened its meaning to include *unusual* or *rare,* writers should be careful about using it too loosely. There could be little harm in describing a unique opportunity, for instance. On the other hand, unique should not be qualified with an adverb, as *very unique, most unique.* And, in most cases, the words *rare, exceptional, unusual,* or *remarkable* can be used, thus obviating the need for *unique.*

UNTIL SUCH TIME AS. Another inflated phrase to be avoided—it simply means *until.* For example: "The pressure must be applied *until such time as* the gauge registers 'Full' " means, "The pressure must be applied *until* the gauge registers 'Full.' "

UTILIZE. A rather inflated verb that actually means to *make useful* or *derive utility from.* According to Webster's *Third New International Dictionary, utilize* "suggests the discovery of a new, profitable, or practical use for something."

Somehow, *utilize* has become synonymous with *use,* which is perfectly understandable but often a bit pompous. To utilize a screwdriver to drive home a one-inch screw, for example, is a bit much.

See also **Inflated Expressions.**

V

VARY, VERY. The difference between these words should be obvious, but occasionally it is not. For example:

1. The answer is now *very* (adverb) clear.
2. The results *vary* (verb) from poor to *very* superior.

VERB. Verbs indicate action or express a state of being. Writers should be aware of several important concepts concerning verbs.

1. *Classification of verbs*
   Verbs may be classified as (a) transitive or intransitive, (b) active or passive, and (c) regular or irregular.
   a. A *transitive* verb requires a direct object to complete the meaning; an intransitive verb requires no object:

      Transitive: I read the *book* (direct object).
      Intransitive: I cannot concentrate (no object).

   b. An *active* verb stresses the actor; i.e., when the subject of the verb carries out the action, the verb is in the active voice. Conversely, when the subject is acted upon, the verb is in the *passive* voice:

      Active: The car struck a wall.
      Passive: The wall was struck by a car.

   Thus, when you want to stress the actor, use the active voice; when the action is the important element, use the passive voice. For a complete discussion of active and passive voices, see **Voice.**

   One other verb form is the *linking* verb, which itself has little meaning but functions to link (connect) the subject and predicate. The most common linking verb is *be* in its various forms:

      (1) John *is* a good student.
      (2) The night *was* cold.
      (3) They *were* late for class.

   c. Other verbs may be used as linking verbs:

      (1) She *became* ill.
      (2) That steak *looks* delicious.
      (3) This coffee *tastes* wonderful.

   d. *Regular and irregular verbs*
      Regular verbs are those whose past tense and past participle are formed by adding *ed* or *d* to the present form. *Regular* verbs, then, have only two forms to denote the three principal parts (present,

past, and past participle): for example, *play* (present tense), *played* (past tense), *played* (past participle).

Some *irregular* verbs have three distinct principal forms, although others have only two, all of which can cause considerable confusion for writers. For instance, the verb *bite* has three distinct forms (*bite, bit, bitten*), whereas *lead* has just two forms (*lead, led, led*). The verb *set,* incidentally, has but one form for all three principal parts (*set, set, set*).

With some irregular verbs, moreover, two forms for the past tense—and in some cases for the past participle—are available, either of which is acceptable. These are indicated by an asterisk in the list of irregular verbs that follows. Although the list contains many common irregular verbs, it is not exhaustive; writers are advised not to guess at verb forms. When in doubt, consult a modern dictionary.

Present	**Past**	**Past Participle**
begin	began	begun
bite	bit	bitten
blow	blew	blown
bring	brought	brought
buy	bought	bought
choose	chose	chosen
come	came	come
deal	dealt	dealt
dive	*dived or dove	dived
do	did	done
draw	drew	drawn
dream	*dreamed or dreamt	*dreamed or dreamt
drink	drank	drunk
drive	drove	driven
eat	ate	eaten
fall	fell	fallen
flee	fled	fled
fly	flew	flown

freeze	froze	frozen
go	went	gone
grow	grew	grown
know	knew	known
lay	laid	laid
lead	led	led
lend	lent	lent
lie	lay	lain
light	*lit or lighted	*lit or lighted
make	made	made
prove	proved	*proved or proven
ride	rode	ridden
rise	rose	risen
run	ran	run
saw	sawed	*sawed or sawn
see	saw	seen
send	sent	sent
set	set	set
shine	*shone or shined	*shone or shined
show	shown	shown
sing	sang	sung
sink	sank	sunk
sit	sat	sat
sow	sowed	*sowed or sown
stand	stood	stood
swim	swam	swum
swing	swung	swung
throw	threw	thrown
wake	*woke or waked	woken
wear	wore	worn
weave	*wove or weaved	*woven or weaved

win	won	won
wind	wound	wound
write	wrote	written

2. *Tenses of verbs*
   The tense of a verb indicates the time (past, present, or future) or the continuance or completion (called the *perfect*) of the action or state. The present or past tense is indicated by the verb ending; the future tense is indicated by the verb ending and the addition of the auxiliary *will* or *shall* (see **Shall, Will**); the perfect tense is indicated by the participial form and the auxiliary *has, have,* or *had.* The progressive form, which indicates that an action is or was "in progress," is formed by adding *ing* to the present form. For example:

Tense	Ordinary Form	Progressive Form
present	he walks	he is walking
past	he walked	he was walking
future	he will walk	he will be walking
perfect	he has walked	he has been walking
past perfect	he had walked	he had been walking
future perfect	he will have walked	he will have been walking

See also **Participle.**

VERBAL NOUN. See **Gerund.**

VICE (as a prefix). See **Prefixes and Suffixes.**

VICE, VISE. Here we are concerned with two nouns—not the prefix that denotes second in command, as in vice-president. Simply remember that most of us have at least one *vice,* or a small personal foible or two, and that the carpenter's tool is the *vise.*

VICE VERSA. An expression meaning *the order being changed* (from that of a preceding statement). It is occasionally useful in statements such as, "Follow the instructions in the chart to convert United States units to metric units and vice versa." (Be careful to spell it correctly; it is not italicized.)

VOICE. When the subject of the verb carries out the action, the verb is said to be in the *active* voice. Conversely, when the subject is acted upon, the verb is called *passive*. Writers are often advised to shun the passive voice in favor of the active, which usually results in a stronger sentence. Consider the difference between these examples:

Passive: A report was assigned by the instructor.
Active: The instructor assigned a report.

Passive: Two more weeks are needed by us to finish the work.
Active: We need two more weeks to finish the work.

Passive (with apologies to Thoreau): Lives of quiet desperation are led by the mass of men.
Active (and as Thoreau wrote it): The mass of men lead lives of quiet desperation.

All three of those passive statements are weak and pompous; the active versions are both direct and natural sounding. On the other hand, the passive voice does have a definite place and in some cases is preferable to the active voice. Consider, for example, this sentence describing an "action": "After casting, the valves are washed, dried, and packed in light grease." As you can see, the action is the important thing here, and the use of passive voice is natural. To use the active voice, you would have to write, ". . . the washer washes them, the drier dries them, and the packer packs them in light grease." In this instance, you are not concerned with the actors, and the use of the active voice would result in a rather fatuous and redundant statement.

However, the active voice has yet another virtue: it almost always reveals who did what. When you conceal the actors in such constructions as *it is believed* or *it is estimated,* readers are left wondering who the unnamed believers or estimators may be. For instance:

Passive and vague: It is estimated that the project will be finished in six months.
Active and specific: The construction manager has estimated that the project will be finished in six months.

As you can see, the active voice attributes that statement to a definite source.

In general, you can resolve the choice of active or passive voice by considering what will appear more natural or informative to your readers. Keep in mind: to stress the actor, use the active voice; to stress the action, use the passive voice.

See also **Sentence Types and Construction.**

W

WHENCE. Although somewhat obsolete, *whence* is occasionally used as a conjunction meaning *from where* or *from what place.* "He returned whence he came." Note, however, that *whence* means *from where;* thus *from whence* is redundant.

WHETHER. *Whether* is a subordinating conjunction introducing a condition or expression of doubt; *whether* is frequently used with the correlative *or.* For instance:

1. He is wondering *whether* a new home *or* an older one would be the better investment.
2. The important question is *whether* it will rain tomorrow.

Note: Since *whether* implies a condition, *whether or not* is somewhat redundant. In sentence 2 above, *or not* is implied by *whether.*

WHICH, THAT. See **Pronoun; Restrictive and Nonrestrictive Modifiers.**

WHO, WHOM, WHOEVER, WHOMEVER. Although *whom* and *whomever* are the objective forms of *who* and *whoever,* respectively, they are becoming increasingly rare in modern writing. *Whom* is occasionally still seen in such constructions as "They couldn't decide whom to vote for," or "You want to speak to whom?"

In the past, particularly when ending a sentence with a preposition was tantamount to committing mayhem, writers would prepare such engaging statements as "Jones is the man for whom I intend to vote." Today, we can write with impunity, "Jones is the man I intend to vote for." There, of course, the question of *whom* is no longer relevant.

In many cases, *who* or *whoever* is the subject of a following clause and thus properly is nominative:

1. Regardless of who is playing, I'm not going to the game tonight.

Although the clause *who is playing* is the object of the preposition *of, who* is the subject of the clause and thus is nominative.

2. I'll give this old car to whoever will take it.

In this case, *whoever* is the subject of the objective clause *whoever will take it* and is properly nominative.

WHO'S, WHOSE. *Who's* is the contraction meaning *who is. Whose* (no apostrophe) is the possessive form of *who.* For example:

1. *Who's* (who is) on duty tonight?
2. *Whose* (possessive) books are these?

Although it normally refers to a person, *whose* is frequently used with inanimate objects to avoid the awkward *of which:*

The store, *whose* shelves sagged with old books of every description, was almost deserted. (Instead of: the shelves *of which* sagged. . . .)

WISE AS A SUFFIX. I once asked a group of students to write their reasons for studying technical writing. One reply was a classic: "To improve my status, jobwise." Not long afterward, I read a similar "classic" statement describing an automobile show: "Carwise, every imaginable type is on display."

The only defensible use of *wise* as a suffix I know of is *lengthwise* (although I have never heard the expression *widthwise*). A word to the "wise": avoid *wise* as a suffix, and say what you mean:

1. If I improve my writing, I will be eligible for promotion.
2. Every type of car is on display.

Y

YOUR, YOU'RE. *Your* is the possessive form of *you; you're* is the contraction meaning *you are:*

1. Don't forget *your* (possessive) books.
2. *You're* (you are) all coming at six o'clock.

Index